T0269221

CAMBRIDGE MONOGRAPHS ON
MATHEMATICAL PHYSICS

General editors: P.V. Landshoff, D.R. Nelson, D.W. Sciama, S. Weinberg

THE STRUCTURE OF THE PROTON

This graduate/research level book describes our present knowledge of protons and neutrons, the particles which make up the nucleus of the atom.

Experiments using high energy electrons, muons and neutrinos reveal the proton as being made up of point-like constituents, quarks. The strong forces which bind the quarks together are described in terms of the modern theory of quantum chromodynamics (QCD), the 'glue' binding the quarks being mediated by new constituents called gluons. Larger and new particle accelerators probe the interactions between quarks and gluons at shorter distances. The understanding of this detailed substructure and of the fundamental forces responsible is one of the keys to unravelling the physics of the structure of matter.

This book will be of interest to all theoretical and experimental particle physicists.

THE STRUCTURE OF THE PROTON

Deep Inelastic Scattering

R. G. ROBERTS

Rutherford Appleton Laboratory

CAMBRIDGE
UNIVERSITY PRESS

CAMBRIDGE UNIVERSITY PRESS
Cambridge, New York, Melbourne, Madrid, Cape Town, Singapore,
São Paulo, Delhi, Dubai, Tokyo, Mexico City

Cambridge University Press
The Edinburgh Building, Cambridge CB2 8RU, UK

Published in the United States of America by
Cambridge University Press, New York

www.cambridge.org
Information on this title: www.cambridge.org/9780521449441

© Cambridge University Press 1990

First published 1990
First paperback edition 1993

A catalogue record for this publication is available from the British Library

Library of Congress Cataloguing in Publication Data

Roberts, R.G.
The structure of the proton : deep inelastic scattering / R. G. Roberts
p.   cm. – (Cambridge monographs on mathematical physics)
Includes bibliographical references (p.   ) and index.
ISBN 0-521-35159-6
1. Protons.   2. Particles (Nuclear physics)   3. Deep inelastic collisions.
4. Field theory (Physics)   I. Title.   II. Series.
QC793.5.P72R63 1990
539.7'212–dc20   90-2292
                                CIP

ISBN 978-0-521-35159-1 Hardback
ISBN 978-0-521-44944-1 Paperback

Contents

Contents

Preface

The description of interactions of particles in terms of the standard model is proving remarkably robust. In this scheme, the electroweak interactions between the leptons and quarks are described by the gauge field theory with broken $SU(2) \times U(1)$ symmetry with the associated gauge bosons of the photon, W and Z, while the strong interactions of quarks are governed by quantum chromodynamics (QCD) with $SU(3)$ symmetry where the gauge boson is the gluon. QCD provides a theoretical framework for formulating the structure of hadrons, in particular that of the proton, in terms of quarks and gluons. This structure is revealed when the proton is probed by a virtual photon or weak current at high energy. This is known as deep inelastic scattering by leptons off a nucleon target and is the subject of this book.

I am indebted to many people for educating me in the subtleties of theory and experiment. Especially I thank Graham Ross, Frank Close, Alan Martin, James Stirling and Roger Phillips on the theoretical and phenomenological front. I have benefitted from discussions with Erwin Gabathuler, Peter Norton and Terry Sloan who with their colleagues provided much of recent excitement on the experimental front. Correspondence with Jan Kwiecinski and Louis Miramontes was a great help. I am grateful to Greg Moley for preparing the text in TeX. Finally I thank Peter Landshoff for continual encouragement.

1

Introduction

Scattering electrons off an atomic nucleus is a very effective way of investigating the electric charge and magnetisation densities of the nucleus. The experiment is shown schematically in fig. 1.1; an electron with four-momentum $k = (E, \mathbf{k})$ scatters off the nucleus at rest and emerges at an angle θ with momentum $k' = (E', \mathbf{k}')$.

The effective probe of the structure within the nucleus is the exchanged virtual photon with momentum $q = k - k'$ with $q^2 = -Q^2$ where $Q^2 > 0$. The resolving power of this probe is the *wavelength* $\hbar/Q$ and so the degree of structure revealed increases with Q^2 which depends on the energy E and the scattering angle θ. In the lab frame, the energy of the virtual photon is $\nu = E - E'$.

For low values of $Q^2 (\sim 0.01 \text{ GeV}^2)$ the nucleus tends to recoil as a whole (coherent scattering) with $\nu = \nu_{el} = Q^2/2M_A$ where M_A is the nuclear mass, or move into an excited state with $\nu = \nu_{el} + (M_{A^*}^2 - M_A^2)/2M_A$. It is useful to use the variable $x_A = Q^2/2M_A\nu$ so that elastic scattering off the nucleus corresponds to $x_A = 1$. As Q^2 increases, the finite size of the nucleus implies a rapid decrease of the elastic cross-section so that by the time Q^2 reaches $\sim 0.1 \text{ GeV}^2$, the most probable process is the scattering off the constituent nucleons, with $\nu = Q^2/2M$ with M being the nucleon mass, or $x_A = M/M_A \simeq 1/A$. The fermi-motion of the nucleons smears out this quasi-elastic peak and those of the nucleon resonances, Δ, N^*.

As the cross-section is now dominated by the scattering off individual nucleons, it is more appropriate to use $x = Q^2/2M\nu$ rather than x_A. As Q^2 moves above 0.5 GeV^2, the form factors of the nucleon and resonances decrease rapidly and we enter the region where the structure of the nucleon is being probed. The proton and neutron are clearly not point-like particles, they have a size typically of ~ 1 fm and their magnetic moments are $\mu_p = 2.79$ and $\mu_n = -1.91$. This departure from Dirac particle

1

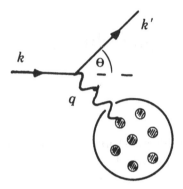

Fig. 1.1   Schematic view of electron nucleus scattering.

magnetic moment values was well known before the advent of
the quark model, such behaviour being thought natural if the
proton structure included states of a nucleon plus one or more
pions.   Nowadays we explain μ_p, μ_n in terms of the point-like
constituent quarks of the proton *(uud)* and neutron *(udd)* with
charges $Q_u = \frac{2}{3}, Q_d = -\frac{1}{3}$ and so have magnetic moments given
by

$$\mu_u = \frac{1}{5}(4\mu_p + \mu_n)\mu_0$$
$$\mu_d = \frac{1}{5}(4\mu_n + \mu_p)\mu_0$$

(1.1)

Here $\mu_0 = \hbar/2Mc$ is the proton's Dirac moment and so (1.1)
gives the masses of the constituent quarks as 330 MeV.   By
analogy with the constituents of the nucleus revealing themselves
at $x_A \simeq 1/A$ we might expect the quark structure to produce
peaks around $x = \frac{1}{3}$.   In practice, we see a smooth continuum, as
Q^2 increases, rising towards $x = 0$ which we can attribute to a *sea*
of quark-antiquark pairs.

   Historically, it was a series of experiments at the Stanford
linear accelerator (SLAC) in the late 1960s on deep (Q^2 *large*)
inelastic scattering off proton and deuterium targets that showed
the nucleon to be made of *hard* point-like objects, or *partons*.
The electrons scattered at large angles thirty times or so more
abundantly than expected (Panofsky 1968).   Furthermore, the
first experiments (Bloom *et al.* 1969, Breidenbach *et al.* 1969)
indicated the famous *scaling* of the structure functions.   These

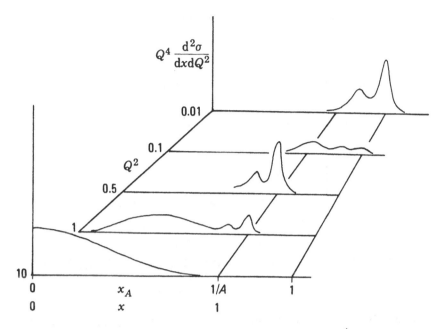

Fig. 1.2 Qualitative features of the cross-section (times Q^4) for electron nucleus scattering.

structure functions W_1 and νW_2, which characterise the features of the scattering other than the photon propagator, were observed to depend only on one scaling variable, x defined above – see fig. 1.3.

This scaling behaviour had been predicted by Bjorken (1969). Starting from current algebra, he argued that, in the limit of $\nu, Q^2 \longrightarrow \infty$, Q^2/ν fixed,

$$\begin{aligned} MW_1(x, Q^2) &\longrightarrow F_1(x) \\ \nu W_2(x, Q^2) &\longrightarrow F_2(x) \end{aligned} \qquad (1.2)$$

That the currents coupling to the photon consisted of spin $\frac{1}{2}$ particles (current quarks) led Callan and Gross (1969) to relate the two scaling structure functions

$$F_2(x) = 2xF_1(x) \qquad (1.3)$$

from which it follows that the cross-section σ_L for longitudinally polarised photons must vanish. It was Feynman (1969) who gave the intuitive picture of the quark parton model (QPM) to understand Bjorken scaling. The proton was regarded as a collection of partons off which the virtual photon scattered

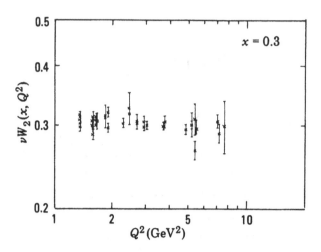

Fig. 1.3 Q^2 dependence of νW_2 for $x = 0.3$ from ep scattering at SLAC.

elastically and the deep inelastic scattering (DIS) cross-section was just the incoherent sum of these individual cross-sections. In the high proton momentum limit $(P \longrightarrow \infty)$ the Bjorken x variable was just the fraction of the proton's momentum carried by the elastically scattered parton, and the size of this cross-section was proportional to the probability $q_i(x)$ of finding a quark of type i and momentum xP in the proton.

By combining the information got by using neutrino and anti-neutrino beams (charged weak current exchange) with the experiments using electrons and muons and by varying the target, the various quark probability distributions could be isolated. The u and d valence quarks were consistent with vanishing at $x = 0$ and $x = 1$ while the sea-quarks tended to populate the small-x region. Furthermore it became clear that the quarks carried only roughly one-half the proton's momentum. The QPM became *the* model to describe all 'hard' processes, i.e. those reactions involving large momenta transferred from leptons (or photons) to hadrons, or from one group of hadrons to another. Thus, dilepton production (Drell and Yan 1971) and large p_T hadroproduction could be computed in terms of the parton distributions extracted from DIS.

The key factor in the QPM explanation of scaling was that, over the short time scale in which the hard scattering took place, the quarks behaved as if they were *free* particles, i.e. no mutual interaction occurring between them. Inevitably, the hunt began for a field theory for quarks which, in the asymptotic limit $Q^2 \longrightarrow \infty$, would describe quarks as *free* particles. This is equivalent to demanding the effective 'charge' in an interacting theory to vanish as smaller and smaller distances are probed. Up to 1973 we were familiar only with theories in which the *reverse* was true: because of the screening of charge at large distances, the coupling in QED becomes slightly smaller at large distances.

Then 't Hooft (1972), Gross and Wilczek (1973) and Politzer (1973) made the breakthrough: *non*-Abelian theories possessed the crucial property of *asymptotic freedom* and, as a result, QCD instantly became the accepted quark field theory and the established theory for strong interactions in general. As in QED there is a screening at large distance of the (colour) charge by quarks and also by the *physical* (i.e. transversely polarised) gluons but this is more than compensated for by an *anti*-screening due to gluons with longitudinal polarisation. Conversely as the momentum scale $Q^2 \longrightarrow \infty$, the effective coupling between quarks and gluons tends to vanish.

In the QCD Lagrangian the basic coupling g is dimensionless. When one computes the loop corrections to the quark gluon coupling, the result diverges and the renormalisation procedure is forced to introduce a scale into the definition of the effective coupling. This coupling sums the leading logarithms of the loop corrections and 'runs', i.e. depends on the scale relevant to the QCD subprocess. The effective coupling is written, to leading order, as

$$\alpha_s(Q^2) = \frac{4\pi}{\beta_0 \ln(Q^2/\Lambda^2)} \qquad (1.4)$$

where $\beta_0 > 0$ and Λ is the scale introduced by renormalisation. Thus $\alpha_s(Q^2) \to 0$ as $Q^2 \to \infty$, making the quarks asymptotically free. We expect the magnitude of Λ to be set by the size of a typical hadron and taking the proton radius ~ 1 fm we get $\Lambda \sim 200$ MeV which is close to experimental estimates.

The QPM with its features of exact scaling and $\sigma_L = 0$ can be regarded as the zeroth approximation of perturbative QCD. The latter can describe the approach to the asymptotic limit, i.e. it can make reliable predictions in the region where α_s is small.

A given operator typically has a high energy behaviour $\sim \alpha_s^d$ where the exponent d (anomalous dimension of the operator) is calculable in the theory. Thus in QCD, the structure functions violate scaling logarithmically, σ_L is still small – proportional to α_s. Today we have a vast amount of experimental data on processes where perturbative QCD can claim to describe the strong interaction physics – $e^+e^- \rightarrow$ hadrons, heavy flavour production, dilepton production, jet production, direct photon production, W, Z production, hadron production in DIS and so on. Estimates of α_s and Λ do not vary too much between the different processes (Duke and Roberts, 1985) but our detailed knowledge of the quark distributions in the proton still comes only from DIS.

What can we say about the glue in the proton? Although the gluon distribution is a crucial component in DIS it enters only *in*directly – through the mixing with the $q\bar{q}$ sea – and so it has been difficult to measure its properties, other than populating the small-x region and carrying half the total momentum. The next few years will almost certainly see a far more precise determination of this constituent of the nucleon. The electron-proton collider at Hamburg (HERA) will probe not only DIS at very high Q^2 (up to 10^6 GeV^2) but will also penetrate to much smaller values of x (down to 10^{-3}) where the dynamics of DIS becomes dominated by the gluon. We shall, at last, be able to see directly the implications of gluon distributions which may be quite singular. In the USA the SSC facility will be able to probe even smaller values of x (down to 10^{-4}) in proton-proton reactions where the gluon-gluon interaction may be the dominant subprocess.

DIS is normally thought of as a testing ground for QCD only in the perturbative sector. This is because the high Q^2 probe of the virtual photon assures us that it is the short distance region which is relevant. However, insofar that we believe that the forces responsible for binding nucleons in the nucleus must ultimately be described in terms of quark and gluon interactions, then differences in DIS off bound and free nucleons must be some sort of measure of QCD forces operating at long-range. That there was any difference came as a surprise – the 'EMC effect' (Aubert *et al.* 1982) – but the modification of the quark structure in a bound proton compared to that in a free proton is now regarded as natural, even if the various explanations put forward might seem to have little physics in common.

In this book we look at the proton's structure as revealed by DIS experiments. As we shall see this corresponds to the region where the relevant momenta are close to the light cone. Thus it is not the structure described by the form factor of the proton that we shall study but rather the high Q^2 limit of the proton structure. The theoretical framework of the operator product expansion allows a separation of physical quantities into factors corresponding to the short and long distance physics. Perturbative QCD is the ideal tool to apply to the former but non-perturbative techniques must be called upon to attempt any kind of description of the x dependence of the structure functions. Here progress has been very limited. Only recently (Martinelli and Sachrajda 1989) has there been a serious assault on computing the moments of the valence quark structure functions for the proton using the powerful machinery of lattice gauge theory.

2

Structure functions

2.1 Kinematics and variables

We want to compute the cross-section for deep inelastic scattering (DIS) to lowest order in weak or electromagnetic interactions. A lepton with momentum k scatters off a nucleon of mass M with the exchange of a photon or Z^0 or $W^{\pm}$ with momentum q as shown in fig. 2.1.

The target nucleon has momentum p and the momentum of the hadronic system X is p_X. We define the invariants

$$
\begin{aligned}
q^2 &= (k - k')^2 = -Q^2, \quad Q^2 > 0 \\
s &= (p + k)^2 \\
W^2 &= p_X^2 \\
\nu &= \frac{p \cdot q}{M} = \frac{1}{2M}(W^2 + Q^2 - M^2)
\end{aligned}
\tag{2.1}
$$

and we neglect the mass of the leptons. It is also useful to define scaling variables

$$
x = \frac{-(k - k')^2}{2p \cdot (k - k')}, \qquad y = \frac{p \cdot (k - k')}{p \cdot k}
\tag{2.2}
$$

so that

$$
\begin{aligned}
x &= \frac{Q^2}{2M\nu} = \frac{Q^2}{W^2 + Q^2 - M^2} \\
y &= \frac{2M\nu}{(s - M^2)} = \frac{W^2 + Q^2 - M^2}{(s - M^2)}
\end{aligned}
\tag{2.3}
$$

At high energies, $s \gg M^2$ and in the Bjorken limit we are interested in large Q^2, ν such that x is finite.

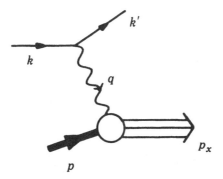

Fig. 2.1   Lowest order graph for deep inelastic scattering.

2.2   Electromagnetic interaction

Consider an electron (or muon) scattering off a nucleon with an electron (or muon) detected at the top vertex in fig. 2.1. Thus the neutral current cross-section involves γ and Z^0 exchange as well as the interference between the two. For $Q^2 < 10^3$ GeV^2 the electromagnetic interaction dominates and, for the moment, we shall consider just the photon exchange. The inclusion of the weak neutral contribution appears in chapter 3.

2.2.1   Structure functions

The amplitude corresponding to fig. 2.1 is

$$T = e^2 \bar{u}(k', \lambda') \gamma^\mu u(k, \lambda) \, \frac{1}{q^2} \, <X|J_\mu^{em}(0)|p, \sigma > \qquad (2.4)$$

Summing over hadronic states, the unpolarised cross-section is

$$d\sigma = \frac{1}{\text{flux}} \frac{d^3 k'}{2k_0'} \frac{1}{4} \sum_{\sigma\lambda\lambda'} |T|^2 \qquad (2.5)$$

which gives

$$k_0' \frac{d\sigma}{d^3 k'} = \frac{2M}{(s - M^2)} \frac{\alpha^2}{Q^4} \, l^{\mu\nu} \, W_{\mu\nu} \qquad (2.6)$$

where the leptonic tensor is

$$l_{\mu\nu}(k, k') = \frac{1}{2} \, \mathrm{Tr}(\not{k}'\gamma_\mu \not{k}\gamma_\nu)$$
$$= 2(k_\mu k'_\nu + k'_\mu k_\nu - \frac{1}{2} \, Q^2 g_{\mu\nu}) \tag{2.7}$$

and the hadronic tensor is

$$W_{\mu\nu}(p, q) = \frac{1}{4M} \sum_\sigma \int \frac{\mathrm{d}^4\xi}{2\pi} e^{iq\cdot\xi} < p, \sigma|[J_\mu^{em}(\xi), J_\nu^{em}(0)]|p, \sigma > \tag{2.8}$$

Because J_μ^{em} is conserved we have $q^\mu W_{\mu\nu} = 0$ and, if we need only unpolarised cross-sections, only terms symmetric in μ, ν appear. Furthermore, using parity conservation and T invariance of the strong and electromagnetic interactions, we can express $W_{\mu\nu}$ in terms of two independent *structure functions*, W_1 and W_2, which are real since J_μ^{em} is hermitian,

$$W_{\mu\nu}(p, q) = - W_1(g_{\mu\nu} + \frac{1}{Q^2}q_\mu q_\nu)$$
$$+ \frac{1}{M^2}W_2(p_\mu + \frac{p \cdot q}{Q^2}q_\mu)(p_\nu + \frac{p \cdot q}{Q^2}q_\nu) \tag{2.9}$$

Using (2.8) and (2.9), the product of the leptonic and hadronic tensors can be expressed in terms of Q^2 and W^2,

$$l^{\mu\nu}W_{\mu\nu} = 2W_1Q^2 + \frac{1}{M^2}W_2[(s-M^2)(s-W^2-Q^2)-M^2Q^2] \tag{2.10}$$

and, since

$$k'_0\frac{\mathrm{d}\sigma}{\mathrm{d}^3k'} = \frac{(s - M^2)}{\pi} \frac{\mathrm{d}^2\sigma}{\mathrm{d}Q^2\mathrm{d}W^2} \tag{2.11}$$

then we get

$$\frac{\mathrm{d}^2\sigma}{\mathrm{d}Q^2\mathrm{d}W^2} = \frac{2\pi\alpha^2 M}{(s - M^2)^2 Q^2}\left[2W_1(W^2, Q^2)\right.$$
$$\left. + W_2(W^2, Q^2)\left\{\frac{(s - M^2)(s - W^2 - Q^2)}{M^2Q^2} - 1\right\}\right] \tag{2.12}$$

or, in terms of ν, Q^2

$$\frac{d^2\sigma}{d\nu dQ^2} = \frac{4\pi\alpha^2 M^2}{(s-M^2)^2 Q^2}\left[2W_1(\nu, Q^2)\right.$$
$$\left.+W_2(\nu, Q^2)\left\{\frac{(s-M^2)(s-M^2-2M\nu)}{M^2 Q^2}-1\right\}\right] \quad (2.13)$$

or, in terms of x, y

$$\frac{d^2\sigma}{dx dy} = \frac{2\pi\alpha^2 M}{(s-M^2)x}\left[2W_1(x, y)\right.$$
$$\left.+W_2(x, y)\left\{\frac{(s-M^2)}{M^2}\frac{(1-y)}{xy}-1\right\}\right] \quad (2.14)$$

The cross-section for an electron scattering in a Coulomb field, Mott-scattering, is

$$\left.\frac{d\sigma}{dQ^2}\right|_{\text{Mott}} = \frac{4\pi\alpha^2}{Q^4}\left[\frac{(s-M^2)(s-W^2-Q^2)}{M^2 Q^2}-1\right] \quad (2.15)$$

and so (2.12) gives

$$\frac{d^2\sigma^{\text{DIS}}}{dQ^2 dW^2} = \frac{1}{2M}\left[2W_1(W^2, Q^2)\left\{\frac{(s-M^2)(s-W^2-Q^2)}{M^2 Q^2}-1\right\}^{-1}\right.$$
$$\left.+W_2(W^2, Q^2)\right]\left.\frac{d\sigma}{dQ^2}\right|_{\text{Mott}} \quad (2.16)$$

Expressions (2.12) – (2.14) for the unpolarised cross-section involve only Lorentz scalars and are therefore suitable for computing the cross-section in any particular frame. For example we can use (2.14) to compute the cross-section in the target (nucleon) rest frame in terms of the incident and final electron energies E, E' and the electron scattering angle θ. For we have

$$E = \frac{(s-M^2)}{2M}, \quad E' = \frac{(s-M^2)}{2M}(1-y),$$
$$\sin^2\frac{\theta}{2} = \frac{M^2 xy}{(s-M^2)(1-y)}, \quad \frac{d\Omega dE'}{dx dy} = \frac{2\pi My}{(1-y)} \quad (2.17)$$

to give

$$\frac{d^2\sigma}{d\Omega dE'} = \frac{\alpha^2}{4E^2 \sin^4\frac{\theta}{2}}\left[2W_1\sin^2\frac{\theta}{2} + W_2\cos^2\frac{\theta}{2}\right] \quad (2.18)$$

$$= \left[2W_1\tan^2\frac{\theta}{2} + W_2\right]\left.\frac{d\sigma}{d\Omega}\right|_{\text{Mott}} \quad (2.19)$$

Note, from (2.17), that in this frame the energy of the virtual photon, $E - E'$ is simply ν. Similarly we can compute in the HERA frame, i.e. in the lab frame for an electron of energy E colliding with a proton of energy E_p with E', θ the energy and direction of the scattered electron. In this frame we have

$$(1 - y) = \frac{E'}{E} \cos^2 \frac{\theta}{2}, \quad xy = \frac{E'}{E_p} \sin^2 \frac{\theta}{2}$$

$$\frac{d\Omega dE'}{dxdy} = \frac{4\pi E \sin^2 \frac{\theta}{2}}{x}$$

(2.20)

to give

$$\frac{d^2\sigma}{d\Omega dE'} = \frac{\alpha^2 M}{8E^2 E_p \sin^4 \frac{\theta}{2}} \left[2W_1 \sin^2 \frac{\theta}{2} + W_2 \frac{4E_p^2}{M^2} \cos^2 \frac{\theta}{2} \right] \quad (2.21)$$

2.2.2   Elastic contribution

Although we are primarily interested in large values for the invariant mass W of the hadronic state, we can compute the contribution to the structure functions from the elastic cross-section. This quantity is described in terms of the electric and magnetic form factors of the nucleon. We have

$$W_1^{el}(W^2, Q^2) = \delta(W^2 - M^2) \frac{Q^2}{2M} G_M^2(Q^2)$$

$$W_2^{el}(W^2, Q^2) = \delta(W^2 - M^2) 2M\mathcal{G}(Q^2)$$

(2.22)

where

$$\mathcal{G}(Q^2) = \left[G_E^2(Q^2) + \frac{Q^2}{4M^2} G_M^2(Q^2) \right] \left(1 + \frac{Q^2}{4M^2} \right)^{-1} \quad (2.23)$$

The expressions (2.22), when inserted into (2.12) give the standard Rosenbluth formula

$$\frac{d\sigma^{el}}{dQ^2} = \frac{2\pi\alpha^2}{(s - M^2)^2} \left[G_M^2(Q^2) + \frac{2\{(s - M^2)^2 - sQ^2\}}{Q^4} \mathcal{G}(Q^2) \right]$$

(2.24)

Since $G_{E,M}(Q^2) \sim (Q^2)^{-2}$ as $Q^2 \to \infty$, clearly the elastic contributions can be neglected at large Q^2. Nevertheless, analysis of experimental data at $Q^2 < 5$ GeV^2 has to include these contributions.

2.2.3 Cross-sections for virtual photons

The structure functions W_1 and W_2 can be related to the absorption cross-sections for virtual transverse and longitudinal photons, σ_T and σ_0. We take standard photon polarisation vectors e_ν satisfying $e \cdot q = 0$, $e_T^2 = -1$, $e_0^2 = 1$ and take σ_T, σ_0 proportional to $e_T^\mu W_{\mu\nu} e_T^\nu$ and $e_0^\mu W_{\mu\nu} e_0^\nu$. This gives

$$\sigma_T = 4\pi\alpha^2 W_1/\text{flux}$$

$$\sigma_0 + \sigma_T = 4\pi\alpha^2(1 + \frac{\nu^2}{Q^2})W_2/\text{flux} \qquad (2.25)$$

There are differing choices for the appropriate definition of the flux but the usual one is

$$\text{flux} = \frac{1}{2M}\sqrt{\lambda}(W^2, q^2, M^2) = \frac{1}{2M}\sqrt{[(W^2 + Q^2 - M^2) + 4M^2Q^2]}$$
$$= \sqrt{\nu^2 + Q^2}$$

$$(2.26)$$

So we get

$$W_1(\nu, Q^2) = \frac{\sqrt{\nu^2 + Q^2}}{4\pi\alpha^2}\sigma_T$$

$$W_2(\nu, Q^2) = \frac{Q^2}{4\pi\alpha^2\sqrt{\nu^2 + Q^2}}(\sigma_0 + \sigma_T) \qquad (2.27)$$

We can also define a *longitudinal* structure function W_L to accompany the purely *transverse* $W_T = W_1$ by

$$W_L(\nu, Q^2) = \frac{\sqrt{\nu^2 + Q^2}}{4\pi\alpha^2}\sigma_0$$
$$= (1 + \frac{\nu^2}{Q^2})W_2(\nu, Q^2) - W_1(\nu, Q^2)$$

$$(2.28)$$

The ratio of longitudinal to transverse cross-sections is denoted by $R = \sigma_0/\sigma_T$ so that we have

$$\frac{W_1(\nu, Q^2)}{W_2(\nu, Q^2)} = (1 + R)^{-1}(1 + \frac{\nu^2}{Q^2}) \qquad (2.29)$$

From (2.13) and (2.29) we see that, if we wish to define a relative longitudinal polarisation vector ϵ of the virtual photon such that the cross-section is proportional to $\sigma_T + \epsilon\sigma_0$ with $0 \le \epsilon \le 1$, then

$$\epsilon^{-1} = 1 + 2(1 + \frac{\nu^2}{Q^2})\left[\frac{(s - M^2)(s - 2M\nu - M^2)}{M^2Q^2} - 1\right]^{-1} \qquad (2.30)$$

We can then write

$$\frac{d^2\sigma}{d\nu dQ^2} \bigg/ \frac{d\sigma}{dQ^2}\bigg|_{\text{Mott}} = \frac{1}{2M}\left[\frac{\epsilon^{-1}+R}{1+R}\right]W_2(\nu, Q^2) \qquad (2.31)$$

and

$$\frac{d^2\sigma}{dQ^2 dW^2} = \frac{\alpha}{\pi}\frac{M\sqrt{\nu^2+Q^2}}{Q^2(s-M^2)^2}\frac{1}{1-\epsilon}(\sigma_T + \epsilon\sigma_0) \qquad (2.32)$$

$$= \frac{\alpha}{\pi}\frac{M\sqrt{\nu^2+Q^2}}{Q^2(s-M^2)^2}\frac{\sigma_T}{1-\epsilon}(1+\epsilon R) \qquad (2.33)$$

The positivity of σ_T, σ_0 leads to the inequalities

$$0 \le W_1 \le (1+\frac{\nu^2}{Q^2})W_2 \qquad (2.34)$$

The first inequality becomes an equality when σ_T vanishes and the second when σ_0 vanishes. We shall see that in the quark parton model σ_0 is expected to vanish in the Bjorken limit giving the relation

$$2xMW_1 = \nu W_2 \qquad (2.35)$$

2.3   Weak charged current interaction

Consider the neutrino induced reactions $\nu N \to \mu^- X$ and $\bar{\nu} N \to \mu^+ X$ or the charged current processes measured at HERA, $e^- p \to \nu X$, and $e^+ p \to \bar{\nu} X$. Here the nucleon is being probed by a virtual $W^\pm$.

2.3.1   Structure functions

The amplitude corresponding to fig. 2.1 is

$$T = \sqrt{2}G_F \bar{u}(k')\gamma^\mu(1-\gamma_5)u(k)(1+Q^2/M_W^2)^{-1} < X|J_\mu^{\text{weak}}(0)|p,\sigma > \qquad (2.36)$$

where $G_F = 10^{-5}/M^2$, $M_W = 82$ GeV. Again we write

$$d\sigma = \frac{1}{\text{flux}}\frac{d^3k'}{2k_0'}\frac{1}{4}\sum_\sigma |T|^2 \qquad (2.37)$$

giving

$$k_0'\frac{d\sigma}{d^3k'} = \frac{M}{(s-M^2)}\frac{G_F^2}{(4\pi)^2(1+Q^2/M_W^2)^2}l^{\mu\nu}W_{\mu\nu} \qquad (2.38)$$

Now the leptonic tensor is

$$
\begin{aligned}
l_{\mu\nu} &= \bar{u}(k)\gamma_\mu(1-\gamma_5)u(k') \cdot \bar{u}(k')\gamma_\nu(1-\gamma_5)u(k) \\
&= 8[k_\mu k'_\nu + k_\nu k'_\mu - g_{\mu\nu}k \cdot k' + i\epsilon_{\mu\nu\rho\sigma}k^\rho k'^\sigma]
\end{aligned}
\tag{2.39}
$$

This time $q^\mu W_{\mu\nu} \neq 0$ since J_μ^{weak} is not conserved and $W_{\mu\nu}$ is first expressed in terms of six structure functions. But CP invariance removes the $(p_\mu q_\nu - p_\nu q_\mu)$ term. Also $q^\mu l_{\mu\nu}$ gives terms proportional to the lepton mass and $(p_\mu q_\nu + p_\nu q_\mu)$ leads to terms like m_{lept}/E while the $q_\mu q_\nu$ terms give $(m_{\text{lept}}/E)^2$, so we need only retain three terms in the hadronic tensor,

$$
\begin{aligned}
W_{\mu\nu} = &- W_1(g_{\mu\nu} + \frac{1}{Q^2}q_\mu q_\nu) \\
&+ \frac{1}{M^2}W_2(p_\mu + \frac{p \cdot q}{Q^2}q_\mu)(p_\nu + \frac{p \cdot q}{Q^2}q_\nu) \\
&- \frac{i}{M}W_3\epsilon_{\mu\nu\rho\sigma}p^\rho q^\sigma
\end{aligned}
\tag{2.40}
$$

Using (2.39), (2.40) and inserting the product of the leptonic and hadronic tensors into (2.38) we get

$$
\begin{aligned}
\frac{d\sigma^\pm}{dxdy} = &\frac{G_F^2(s - M^2)}{2\pi(1 + Q^2/M_W^2)^2}\Big[xy^2 MW_1(x,y) \\
&+ \Big\{1 - y - \frac{xyM^2}{(s-M^2)}\Big\}\nu W_2(x,y) \pm xy(1-\tfrac{1}{2}y)\nu W_3(x,y)\Big]
\end{aligned}
\tag{2.41}
$$

where the $\pm$ refers to $W^\pm$ for the charged current. The last term in (2.41) corresponds to the interference between the vector and axial-vector currents. Again we can compute (2.41) in any frame, e.g. in the target (nucleon) rest frame. Using (2.17) we get

$$
\begin{aligned}
\frac{d^2\sigma^\pm}{d\Omega dE'} = &\frac{G_F^2 E'^2}{2\pi^2(1 + Q^2/M_W^2)^2} \\
&\times \Big[2W_1 \sin^2\frac{\theta}{2} + W_2 \cos^2\frac{\theta}{2} \mp \frac{(E + E')}{M}W_3 \sin^2\frac{\theta}{2}\Big]
\end{aligned}
\tag{2.42}
$$

2.3.2   Elastic contribution

As one might expect, the elastic contribution ($x = 1$) involves the weak dipole form factors of the nucleon F_A and F_V. The result which reproduces the formula for $\nu N \to \mu^- N$ is

$$W_1^{el}(x, Q^2) = \delta(x-1)\frac{M}{Q^2}\left[\lambda^2(1 + \frac{Q^2}{4M^2})F_A^2 + \beta^2\frac{Q^2}{4M^2}F_V^2\right]$$

$$W_2^{el}(x, Q^2) = \delta(x-1)\frac{M}{Q^2}\left[\lambda^2 F_A^2 + \frac{(1 + \beta^2 Q^2/4M^2)}{(1 + Q^2/4M^2)}F_V^2\right]$$

$$W_3^{el}(x, Q^2) = \delta(x-1)\frac{M}{Q^2}\left[2\lambda\beta F_A F_V\right]$$

(2.43)

where $\lambda = G_A/G_V \simeq 1.26$, $\beta = \mu_p - \mu_n = 4.71$. Note that W_3^{el} explicitly displays the interference between vector and axial-vector contributions.

2.3.3   Cross-sections for virtual W's

Just as in the electromagnetic case, we can define polarised cross-sections for the exchanged boson and relate these to the structure functions. Let us call σ_L, σ_R and σ_0 the left-, right- transverse and longitudinal cross-sections for the W, then proceeding as in the electromagnetic case we derive the relations

$$W_1 = \frac{\sqrt{\nu^2 + Q^2}}{\sqrt{2}\pi G_F M_W^2}(\sigma_L + \sigma_R)$$

$$W_2 = \frac{1}{\sqrt{2}\pi G_F M_W^2}\frac{Q^2}{\sqrt{\nu^2 + Q^2}}[\sigma_L + \sigma_R - 2\sigma_0] \qquad (2.44)$$

$$W_3 = \frac{2M}{\sqrt{2}\pi G_F M_W^2}(\sigma_L - \sigma_R)$$

Again the structure functions satisfy inequalities as a result of the positivity of the cross-sections. We get

$$\frac{\sqrt{\nu^2 + Q^2}}{2M}W_3 \le W_1 \le (1 + \frac{\nu^2}{Q^2})W_2 \qquad (2.45)$$

If we suppose that in the limit $Q^2 \to \infty$, x fixed, that the products $Q^2\sigma_i$ scale – i.e. depend only on x for $i = L, R, 0$ then we get

$$MW_1(\nu, Q^2) \longrightarrow F_1(x)$$
$$\nu W_2(\nu, Q^2) \longrightarrow F_2(x) \qquad (2.46)$$
$$\nu W_3(\nu, Q^2) \longrightarrow F_3(x)$$

The $F_i(x)$ are the *scale invariant* structure functions which, from (2.45), satisfy

$$xF_3(x) \leq 2xF_1(x) \leq F_2(x) \qquad (2.47)$$

If chiral invariance were exact in DIS then only $\sigma_L \not\to 0$ as $Q^2 \to \infty$ and the inequalities become equalities,

$$xF_3(x) = 2xF_1(x) = F_2(x) \qquad (2.48)$$

In addition we get the scaling of the longitudinal structure function W_L,

$$2MxW_L(\nu, Q^2) \longrightarrow F_L(x) \qquad (2.49)$$

where

$$F_L(x) = F_2(x) - 2xF_1(x) \qquad (2.50)$$

and so under the conditions for (2.48) to hold we expect $F_L = 0$. In the next chapter we shall see how scaling is a property of the quark-parton model and therefore, in that context, we necessarily have σ_L, F_L vanishing.

2.4   Polarised structure functions

So far we have discussed only unpolarised cross-sections which means that only the symmetric terms in the hadronic tensor $W_{\mu\nu}$ were considered. We now go on to study polarised deep inelastic electroproduction in the region where the weak current can be neglected. We first define a polarisation vector S_μ for the nucleon. If ξ is the unit vector along the direction of the polarisation in the nucleon rest frame, then

$$S_\mu = \left(\frac{\mathbf{p} \cdot \xi}{M}, \; \xi + \frac{(\mathbf{p} \cdot \xi)\mathbf{p}}{M(E + M)} \right) \qquad (2.51)$$

which satisfies $S^2 = -1$ and $S \cdot p = 0$. We introduce two extra structure functions G_1 and G_2 corresponding to the two possible *antisymmetric* contributions to $W_{\mu\nu}$ allowed by P and T invariance,

$$W_{\mu\nu}^{(A)}(q, p, S) = \frac{1}{2}(W_{\mu\nu} - W_{\nu\mu}) = \frac{i}{M^2} \epsilon_{\mu\nu\rho\sigma} q^\rho$$

$$\times \left[S^\sigma(G_1(\nu, Q^2) + \frac{\nu}{M}G_2(\nu, Q^2)) - \frac{1}{M^2}(S \cdot q)p^\sigma G_2(\nu, Q^2) \right] \qquad (2.52)$$

These structure functions can be related to the absorptive cross-sections for virtual photons with projection $\frac{1}{2}$ and $\frac{3}{2}$ of the total spin along the direction of photon momentum. If we call these cross-sections $\sigma_{\frac{1}{2}}, \sigma_{\frac{3}{2}}$ and let σ_I be the interference between the transverse and longitudinal polarisations of the photon, then

$$\sigma_{\frac{3}{2}} = \frac{4\pi\alpha^2}{\sqrt{\nu^2 + Q^2}} \left[W_1 - \frac{\nu}{M^2} G_1 + \frac{Q^2}{M^3} G_2 \right]$$

$$\sigma_{\frac{1}{2}} = \frac{4\pi\alpha^2}{\sqrt{\nu^2 + Q^2}} \left[W_1 + \frac{\nu}{M^2} G_1 - \frac{Q^2}{M^3} G_2 \right] \qquad (2.53)$$

$$\sigma_I = \frac{4\pi\alpha^2}{\sqrt{\nu^2 + Q^2}} \frac{\sqrt{Q^2}}{M^2} \left[G_1 + \frac{\nu}{M} G_2 \right]$$

We can also define asymmetries for the virtual photon scattering by

$$A_1 = \frac{\sigma_{\frac{1}{2}} - \sigma_{\frac{3}{2}}}{\sigma_{\frac{1}{2}} + \sigma_{\frac{3}{2}}} = \frac{M\nu G_1 - Q^2 G_2}{M^3 W_1}$$

$$A_2 = \frac{\sigma_I}{\sigma_T} = \frac{\sqrt{Q^2}}{M^3 W_1} (MG_1 + \nu G_2) \qquad (2.54)$$

where σ_T, given by (2.25), is $(\sigma_{\frac{1}{2}} + \sigma_{\frac{3}{2}})/2$.

The cross-sections for polarised electron-proton scattering can then be straightforwardly computed by contracting $W_{\mu\nu}^{(A)}$ with the lepton tensor $l^{\mu\nu}$ where the latter includes the spin- dependent part,

$$l_{\mu\nu}^{(A)} = 2i\epsilon_{\mu\nu\alpha\beta} k^\alpha q^\beta \qquad (2.55)$$

The difference of cross-sections for a proton polarised parallel or antiparallel to the beam direction is proportional to $l_{(A)}^{\mu\nu} W_{\mu\nu}^{(A)}$. In the proton rest frame with $S = (0,0,0,1)$ we get

$$\frac{d^2\sigma(\uparrow\downarrow - \uparrow\uparrow)}{d\Omega dE'} = \frac{4\alpha^2}{M^3 Q^2} \frac{E'}{E} [MG_1(E + E' \cos\theta) - Q^2 G_2] \quad (2.56)$$

Polarising the target *transversely* to the beam gives another combination,

The longitudinal polarisation asymmetry, A, can be expressed in terms of the asymmetries of (2.54)

$$A = \frac{d\sigma(\uparrow\downarrow - \uparrow\uparrow)}{d\sigma(\uparrow\downarrow + \uparrow\uparrow)} = D[A_1 + \eta A_2] \qquad (2.58)$$

where

$$D = \frac{1 - (1 - y)\epsilon}{1 + \epsilon R}, \quad \eta = \frac{2M\epsilon\sqrt{Q^2}}{s[1 - (1 - y)\epsilon]} \qquad (2.59)$$

ϵ being the relative polarisation given by (2.30). If we again suppose that the quantities $Q^2\sigma_i$ satisfy scaling then we define, from (2.53), the scaling polarised structure functions $g_1(x)$ and $g_2(x)$,

$$\frac{\nu}{M}G_1(\nu, Q^2) \longrightarrow g_1(x), \quad \frac{\nu^2}{M^2}G_2(\nu, Q^2) \longrightarrow g_2(x) \qquad (2.60)$$

and we then see that, in this approximation,

$$A_1 \longrightarrow \frac{g_1(x)}{F_1(x)}, \quad A_2 \longrightarrow 0 \qquad (2.61)$$

In practice, D is very close to 1 so that the measurement of the asymmetry A gives the ratio $g_1(x)/F_1(x)$.

2.5 Extracting structure functions from experiment

In a DIS experiment where an electron (or muon) is the final state lepton, the variables Q^2, x and y can be determined from the energy E' and scattered angle θ of this electron. For example, a neutral current event at HERA, where $s = 4EE_p$, gives (using (2.20))

$$Q^2 = 4EE' \sin^2 \frac{\theta}{2}$$

$$x = \frac{E' \sin^2 \frac{\theta}{2}}{E_p[1 - (E'/E) \sin^2 \frac{\theta}{2}]} \qquad (2.62)$$

$$y = 1 - \frac{E'}{E} \sin^2 \frac{\theta}{2}$$

where the angle θ is measured from the incoming electron beam. However when the final state lepton is a neutrino (or antineutrino) the variables Q^2, x, y must be reconstructed from the measurements of the energy E_J and angle θ_J of the jet of hadrons associated with the current. In this case, the quark parton model interprets this jet as the quark after it has been scattered by the $W^{\pm}$ and we get

$$Q^2 = \frac{E_J^2 \sin^2 \theta_J}{1 - (E_J/E)\sin^2 \frac{\theta_J}{2}}$$

$$y = \frac{E_J}{E} \sin^2 \frac{\theta_J}{2} \qquad (2.63)$$

$$x = \frac{E_J \cos^2 \frac{\theta_J}{2}}{E_p(1 - \frac{E_J}{E}\sin^2 \frac{\theta_J}{2})}$$

So it is certainly easier to extract structure functions when the scattered lepton can be detected. Suppose we were doing this when the target nucleon was at rest and we needed W_1, W_2 at fixed values of ν and Q^2. From (2.18) we see that the differential cross-section must be measured at different angles. Since θ is a function of ν, Q^2 and s then the separation requires different values of the incoming energy. If this is done, then $R = \sigma_L/\sigma_T$ can be extracted at each value of ν and Q^2, using (2.33).

It is very difficult to get precise determinations of R, especially as a function of two variables, e.g. x and Q^2. Most experiments can achieve a 'global' value obtained from averaging over their entire data. Having fixed R, the W_2 structure function can then be directly obtained from the electroproduction cross-section by

$$\nu W_2(x, Q^2) = \left[\frac{\mathrm{d}^2\sigma/\mathrm{d}\Omega\mathrm{d}E'}{(\mathrm{d}\sigma/\mathrm{d}\Omega)_{\mathrm{Mott}}}\right]\left[\frac{\nu\epsilon(1+R)}{1+\epsilon R}\right] \qquad (2.64)$$

and fig. 2.2 shows the result of this exercise, taking $R = 0.21$ and using the entire SLAC data on a proton target at $Q^2 = 3.5$ GeV^2. These beautiful data are precise enough to show, at this low value of Q^2, the Δ and N resonances.

In neutrino induced charged current interactions the separation of the νW_2, νW_3 structure functions is achieved by using both ν and $\bar{\nu}$ beams. Thus, for example, we get from (2.41)

$$x\nu W_3(x, Q^2) = \frac{2\pi(1 + Q^2/M_W^2)^2}{G_F^2 sy(2-y)}\left[\frac{\mathrm{d}^2\sigma^\nu}{\mathrm{d}x\mathrm{d}y} - \frac{\mathrm{d}^2\sigma^{\bar\nu}}{\mathrm{d}x\mathrm{d}y}\right] \qquad (2.65)$$

and fig. 2.3 shows the result of using the wide-band beam data of the CDHSW collaboration (Berge *et al.* 1989) on a target of iron

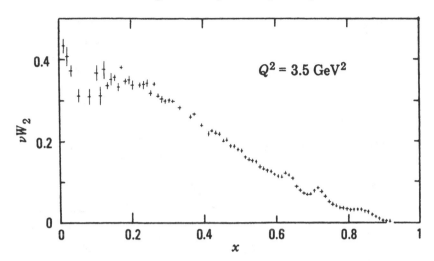

Fig. 2.2 νW_2 structure function of the proton from SLAC electroproduction experiments at $Q^2 = 3.5$ GeV^2. As well as data from E87, E49B, E89-2, E49-A at SLAC, muonproduction data from Fermilab E-98 (small x data with $R = 0.52$) are included. See Poucher *et al.* (1974), Bodek *et al.* (1979), Gordon *et al.* (1979) and Mestayer *et al.* (1983).

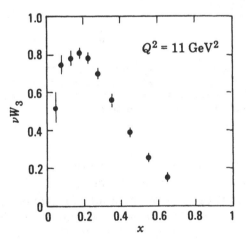

Fig. 2.3 $x\nu W_3$ structure function (xF_3) from $\nu, \bar{\nu}$ beams on iron target at $Q^2 = 11$ GeV^2. Data from the CDHSW collaboration (Berge *et al.* 1989).

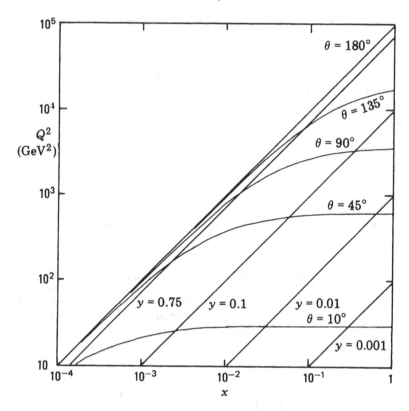

Fig. 2.4 Kinematic region at maximum HERA energies, $E = 30$ GeV and $E_p = 820$ GeV. Note that both Q^2 and x axes are logarithmic. The diagonal lines correspond to fixed values of y and the curves give the values of the angle, θ, of the scattered electron for neutral current DIS.

for $Q^2 = 11$ GeV^2. The structure functions are divided by A (56 in this case) to get the structure function for an isoscalar nucleon.

A large amount of data on structure functions from muonproduction and neutrino DIS from CERN and Fermilab has been measured covering the region $Q^2 < 200$ GeV^2, $x > 0.05$. Even difficult measurements such as the longitudinal structure function F_L and the polarised structure function $g_1(x)$ have now been achieved and we shall show these results when we come to discuss their interpretation in later chapters.

There is a dramatic increase in the kinematic region available for DIS experiments as HERA comes into operation. In fig. 2.4 we show this kinematic region when electrons at 30 GeV collide with protons at 820 GeV which gives $\sqrt{s}$ more than 300 GeV – or equivalent to a fixed target experiment where the incident electron energy is about 50 TeV.

When measuring structure functions the graph of fig. 2.1 is only the lowest order QED contribution. One of the major problems is to make corrections for higher order radiative graphs so that the contribution of fig. 2.1 can be extracted from the data. A large number of graphs give rise to radiative corrections but the largest correction arises from a photon radiated from the electron lines. Clearly this results in a change of the 'observed' value of Q^2 and since the photon propagator in fig. 2.1 gives a factor $1/Q^4$ then the 'correction' can sometimes be almost 100% – at large y for example. In the appendix we discuss these radiative corrections in some detail, presenting, in particular, estimates for the magnitude of the corrections expected at HERA.

3

Quark parton model

We have already mentioned the consequences of the quark parton model (QPM) for the structure functions, scaling and the Callan-Gross relation, for example. The QPM is intuitively described in the classic texts (Feynman 1972, Close 1979) and a clear formal derivation of the QPM is given by Jaffe (1985) for example.

3.1 Light cone dominance and the impulse approximation

Let us consider inelastic electroproduction in the target (nucleon) rest frame with the virtual photon momentum along the z axis, i.e. in fig. 2.1 we have $p = (M, 0, 0, 0)$ and $q = (\nu, 0, 0, -\sqrt{\nu^2 + Q^2})$. In the Bjorken limit, $Q^2 \to \infty$, x fixed we have $q = (\nu, 0, 0, -\nu - Mx)$. It is convenient to define *light-cone* variables $a^{\pm} = (a^0 \pm a^3)/\sqrt{2}$ so that the scalar product $a \cdot b = a^+ b^- + a^- b^+ - \mathbf{a}_T \cdot \mathbf{b}_T$ and then we have

$$q^+ = -Mx/\sqrt{2}, \qquad q^- = (2\nu + Mx)/\sqrt{2} \longrightarrow \sqrt{2}\nu \qquad (3.1)$$

Thus, in the Bjorken limit, $q^- \to \infty$, q^+ is fixed. If we ask, within the context of a space-time description, what sort of distance scales are important for DIS, we turn to the definition (2.8) of the hadronic tensor $W_{\mu\nu}$. We are therefore interested in the size of the scale which dominates ξ, the space-time interval between the points at which the currents $J_\mu(\xi)$ and $J_\mu(0)$ act. The commutator of the currents is weighted by the factor $\exp(iq \cdot \xi)$ where $q \cdot \xi = q^+ \xi^- + q^- \xi^+$. Since $q^- \to \infty$ but $q^+ \to -Mx/\sqrt{2}$ we get

$$\xi^+ \longrightarrow 0 \ , \quad |\xi^-| < \sqrt{2}/Mx \qquad (3.2)$$

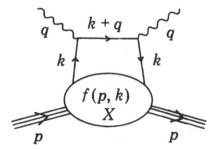

Fig. 3.1   Parton model graph

Thus for $\nu \rightarrow \infty$, we have

$$W_{\mu\nu} \sim \int \mathrm{d}\xi^{-}\, \mathrm{e}^{iq^{+}\xi^{-}} \int \mathrm{d}\xi^{+}\mathrm{d}^{2}\xi_{T}\, \mathrm{e}^{iq^{-}\xi^{+}} < p|J_{\mu}(\xi)J_{\nu}(0)|p > \quad (3.3)$$

and for fixed x, the rapidly oscillating exponential term kills all contributions except the singularities of the integrand. The integrand is indeed singular at $\xi^{+} = 0$ (remembering that causality forces it to vanish for $\xi^{2} < 0$). The result (see more detailed argument by Jaffe 1985) is therefore that the Bjorken limit of DIS is dominated by $\xi^{+} \simeq 0$. Notice that *every* component of ξ (except ξ^{-}) $\rightarrow 0$ in the Bjorken limit. Thus the appropriate description for DIS is not *short* distance physics but rather *light-cone* ($\xi^{2} \rightarrow 0$) *dominated* physics.

Also notice that the relevant time scale $\Delta\xi_{0} = \Delta\xi_{3}$ is $O(1/\sqrt{Q^{2}})$. As a result, the virtual photon probes the target nucleon essentially *'frozen'* on this time scale $\Delta\xi_{0}$ which $\rightarrow 0$ as $Q^{2} \rightarrow \infty$ as such a time scale is far shorter than that which characterises the strong interactions. This is the *impulse approximation* and is one of the crucial assumptions in the QPM.

3.2   Bjorken scaling

In the impulse approximation, DIS is described by the incoherent elastic scattering of the virtual photon off the quarks. The relevant graph is thus the one in fig. 3.1.

Another contribution, the graph of fig. 3.2, is possible but when we study the contributions from figs 3.1 and 3.2 in the context of

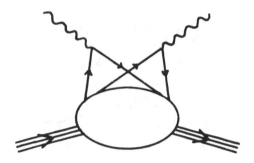

Fig. 3.2   Interference contribution to photon-quark cross-section

perturbative QCD we find that the latter is relatively suppressed by $1/Q^2$ factors.

The diagram of fig. 3.1 allows the hadronic tensor $W_{\mu\nu}$ for unpolarised scattering to be expressed in the form

$$W_{\mu\nu}(q,p) = \sum_i \sum_s \int \mathrm{d}^4k \ f_s^i(p,k) \ w_{\mu\nu}^i(q,k) \ \delta[(k+q)^2] \quad (3.4)$$

where the sums are over the quark flavours, i, and the quark helicities s. In the covariant formulation of the parton model (Landshoff and Polkinghorne 1972) the strong interaction vertex, f_s^i, is a function only of the scalar $p \cdot k$. The $w_{\mu\nu}$ tensor describes the interaction of the virtual photon with the quark of momentum k and in the case of massless on-shell quarks is

$$w_{\mu\nu}^i = \frac{1}{4} e_i^2 \ \mathrm{Tr}[\not{k}\gamma_\mu(\not{k}+\not{q})\gamma_\nu] \quad (3.5)$$

Now let x_i be the fraction of the proton light-cone momentum carried by the quark, i.e. $x_i = k^+/p^+$. Since $k^2 = 2k^+k^- - k_T^2 = 0$, then $k^- = k_T^2/2x_ip^+$ and so $k \cdot q = k^+q^- + k^-q^+ = x_ip^+q^- = x_i \, p \cdot q = x_iM\nu$ and the δ-function becomes

$$\delta[(k+q)^2] = \frac{1}{2M\nu} \ \delta(x_i - x) \quad (3.6)$$

Thus in the QPM, the Bjorken x variable is just the fractional (light-cone) momentum of the struck quark. Evaluating (3.5) and using (3.6) we get

$$W_{\mu\nu}(q,p) = \sum_i e_i^2 \int \frac{\mathrm{d}^4k}{2M\nu} \ [f_+^i(p \cdot k) + f_-^i(p \cdot k)] \, \delta(x_i - x)$$
$$\times [2k_\mu k_\nu + k_\mu q_\nu + q_\mu k_\nu - g_{\mu\nu}k \cdot q] \quad (3.7)$$

To get expressions for the individual structure functions we consider particular values of μ, ν. Take $\mu = \nu = 2$ and go to the proton rest frame, for example. From (2.9) we have, in this case, $W_{22} = W_1$ and using $d^4k = \frac{\pi}{2}\frac{dx}{x}dk^2 dk_T^2$,

$$W_{22} = \sum_i e_i^2 \int \frac{dx dk_T^2}{2x M\nu} \left[f_+^i(p \cdot k) + f_-^i(p \cdot k) \right] \delta(x_i - x)\, x M\nu \quad (3.8)$$

or

$$MW_1(p,q) \longrightarrow F_1(x) = \sum_i e_i^2\, q_i(x) \quad (3.9)$$

where

$$q_i(x) = \frac{\pi}{4} \int dk_T^2 \left[f_+^i(p \cdot k) + f_-^i(p \cdot k) \right] \quad (3.10)$$

is conventionally interpreted as the probability of finding a quark of type i with fraction x of the proton momentum. Note that $2p \cdot k / M^2 \equiv y = x + k_T^2 / x M^2$. Again, taking $\mu = \nu = 0$, we get from (2.9) as $\nu, Q^2 \to \infty$

$$W_{00} = \frac{\nu^2}{MQ^2}\left[-MW_1 + \frac{1}{2x}\nu W_2 \right] \quad (3.11)$$

and from (3.7),

$$W_{00} = \sum_i e_i^2 \int \frac{d^4k}{2M\nu} \left[f_+^i(p \cdot k) + f_-^i(p \cdot k) \right] \delta(x_i - x)\, \nu(2k_0 - xM) \quad (3.12)$$

Since $2k_0 = \sqrt{2}(k^+ + k^-) \to \sqrt{2}xp^+ = xM$ in the rest frame then $W_{00} \to 0$ in the Bjorken limit and so (3.11) gives the Callan-Gross (1969) relation

$$\nu W_2 = 2x M W_1 \quad (3.13)$$

Thus the QPM gives the important results of scaling, given by (3.9) and

$$\nu W_2(p,q) \longrightarrow F_2(x) = \sum_i e_i^2\, xq(x) \quad (3.14)$$

together with

$$F_2(x) = 2x\, F_1(x) \quad (3.15)$$

which, from (2.29), implies that $R = \sigma_0/\sigma_T = 0$. Although we have simplified the kinematics by considering only massless on-shell partons, the results (3.13) – (3.15) follow generally for $k^2 \neq 0$.

The relation (3.15) is a consequence of the spin $\frac{1}{2}$ nature of the partons; spin 0 partons would lead to $W_1 = 0$, $\sigma_T = 0$ and $R = \infty$.

3.3 Quark distributions

It is convenient to express the quark distributions in flavour combinations. Take four flavours and define the combinations:
 flavour singlet:

$$\Sigma(x) = \sum_i (q_i(x) + \bar{q}_i(x)) \qquad (3.16)$$

valence quarks:

$$V(x) = \sum_i (q_i(x) - \bar{q}_i(x))$$
$$= u_v(x) + d_v(x) \qquad (3.17)$$

where $u(x) = u_v(x) + u_s(x)$, and we assume $\bar{u}(x) = u_s(x)$ and similarly for $d(x)$, so that we have
 non-charm sea:

$$S(x) = 2(u_s(x) + d_s(x) + s_s(x)) \qquad (3.18)$$

where the strange quark distribution $s(x) = \bar{s}(x) = s_s(x)$ and finally
 charm sea:

$$C(x) = 2c(x) = c(x) + \bar{c}(x) \qquad (3.19)$$

3.3.1 Electron nucleon scattering via the neutral current

Consider first the electromagnetic structure functions. The convention is to define $u(x), d(x)$ etc. as the distributions of up, down quarks etc. in the *proton*. So the up, down distributions in the neutron are $d(x)$, $u(x)$ respectively. Thus, from (3.13), we get

$$F_2^{ep}(x) = \tfrac{4}{9}[xu(x) + x\bar{u}(x) + xc(x) + x\bar{c}(x)]$$
$$+ \tfrac{1}{9}[xd(x) + x\bar{d}(x) + xs(x) + x\bar{s}(x)] \qquad (3.20)$$

with $F_2^{en}(x)$ obtained from (3.20) with $u \leftrightarrow d$, $\bar{u} \leftrightarrow \bar{d}$. If the non-strange sea is $SU(2)$ symmetric, $\bar{u} = \bar{d}$ and we get

$$F_2^{ep}(x) - F_2^{en}(x) = \tfrac{1}{3}[xu_v(x) - xd_v(x)] \qquad (3.21)$$

i.e. a measurable flavour non-singlet quantity. If we take a target of deuterium then we get the structure function for an isoscalar nucleon, $N = \frac{1}{2}(p+n)$,

$$F_2^{eN}(x) = \tfrac{5}{18}x\Sigma(x) - \tfrac{1}{3}[x\bar{s}(x) - x\bar{c}(x)] \qquad (3.22)$$

and if the second term is small then this quantity becomes a pure flavour singlet combination.

Now let us increase Q^2 so that the weak part of the neutral current must be included, i.e. we have γ-exchange, Z-exchange and γ-Z interference. Consider first a left-handed electron scattering off a proton. The cross-section can be written

$$\frac{d\sigma(e_L^- p)}{dxdQ^2} = \frac{4\pi\alpha^2}{xQ^4}[xy^2F_1(x) + (1-y)F_2(x) + xy(1-\tfrac{1}{2}y)F_3(x)] \qquad (3.23)$$

In terms of the parton distributions, the structure functions are

$$F_2(x) = 2xF_1(x) = \sum_i A_i(Q^2)[xq_i(x) + x\bar{q}_i(x)] \qquad (3.24)$$

$$F_3(x) = \sum_i B_i(Q^2)[q_i(x) - \bar{q}_i(x)] \qquad (3.25)$$

where

$$A_i(Q^2) = e_i^2 - e_i g_{Le}(g_{Li} + g_{Ri})P_Z + \tfrac{1}{2}g_{Le}^2(g_{Li}^2 + g_{Ri}^2)P_Z^2 \qquad (3.26)$$

and

$$B_i(Q^2) = e_i g_{Le}(-g_{Li} + g_{Ri})P_Z - \tfrac{1}{2}g_{Le}^2(-g_{Li}^2 + g_{Ri}^2)P_Z^2 \qquad (3.27)$$

Here e_i, q_{Li}, q_{Ri} are the charge, left- and right-handed weak couplings of a quark of type i and g_{Le}, g_{Re} are the corresponding couplings for the electron. Thus (3.26), (3.27) can be evaluated using

$$P_Z = [(1 + M_Z^2/Q^2)\sin^2 2\theta_W]^{-1}$$

$$g_{Le} = 2\sin^2\theta_W - 1, \quad g_{Re} = 2\sin^2\theta_W$$

$$g_{Li} = 1 - \tfrac{4}{3}\sin^2\theta_W, \quad g_{Ri} = -\tfrac{4}{3}\sin^2\theta_W \qquad \text{for } i = u, c$$

$$= -1 + \tfrac{2}{3}\sin^2\theta_W, \qquad = \tfrac{2}{3}\sin^2\theta_W \qquad \text{for } i = d, s$$

To get the corresponding expression to (3.24) for $e_R^- p$ put $F_3 \to -F_3$ and $g_{Le} \to g_{Re}$. For $e_R^+ p$ put $F_3 \to -F_3$ and for $e_L^+ p$ put $g_{Le} \to g_{Re}$. Of course the pure Z exchange is much smaller than the $\gamma - Z$ interference, as can be seen from fig. 3.3.

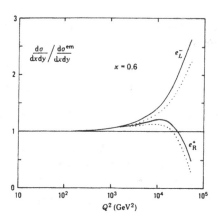

Fig. 3.3 The ratio of the full neutral current cross-section to the electromagnetic contribution at $x = 0.6$ for $e_L^- p \longrightarrow e_L^- X$ and $e_R^+ p \longrightarrow e_R^+ X$ as a function of Q^2. The dotted lines correspond to $\gamma - Z$ interference only.

3.3.2   Neutrino-nucleon scattering via the charged current

For neutrino-quark scattering the hadronic weak current is

$$J_\mu = \bar{u}(x)\gamma_\mu(1 - \gamma_5)[d(x)\cos\theta_c + s(x)\sin\theta_c]$$
$$+ \bar{c}(x)\gamma_\mu(1 - \gamma_5)[s(x)\cos\theta_c - d(x)\sin\theta_c] \qquad (3.28)$$

Take $\nu d \longrightarrow \mu^- u$ for example. Both neutrino and quark are left handed and the total spin $J = 0$ and the scattering is therefore isotropic in the c.o.m. angle θ^*. From (3.28) we get

$$\frac{d\sigma}{d\Omega}(\nu d \longrightarrow \mu^- u) = \frac{G_F^2}{4\pi^2 \hat{s}}\hat{s}^2 \cos^2\theta_c \qquad (3.29)$$

where $\hat{s}, \hat{t}, \hat{u}$ are the relevant invariants in the s, t and u channels. For neutrino-antiquark scattering the spins add to $J = 1$, i.e. the amplitude for $\bar{\nu}u \longrightarrow \mu^+ d$ is got by crossing symmetry $\hat{s} \longleftrightarrow \hat{u}$, with $\hat{u} = -\frac{1}{2}\hat{s}(1 + \cos\theta^*)$, and so

$$\frac{d\sigma}{d\Omega}(\bar{\nu}u \longrightarrow \mu^+ d) = \frac{G_F^2}{4\pi\hat{s}}\frac{1}{4}\hat{s}^2(1 + \cos\theta^*)^2 \cos^2\theta_c \qquad (3.30)$$

The scaling variable $y = \frac{1}{2}(1 - \cos\theta^*)$, so we can write

$$\frac{d\sigma}{dy}(\nu d \longrightarrow \mu^- u) = \frac{G_F^2}{\pi}\hat{s}\,\cos^2\theta_c \qquad (3.31)$$

$$\frac{d\sigma}{dy}(\bar{\nu}u \longrightarrow \mu^+ d) = \frac{G_F^2}{\pi}\hat{s}\,(1 - y)^2 \cos^2\theta_c \qquad (3.32)$$

Table 3.1. Factors appearing in
(3.33).

Sub-process	$f_i(x, \cos\theta_c)$
$\nu d \to \mu^- u$	$xd(x)\cos^2\theta_c$
$\nu s \to \mu^- u$	$xs(x)\sin^2\theta_c$
$\nu s \to \mu^- c$	$xs(x)\cos^2\theta_c$
$\nu d \to \mu^- c$	$xd(x)\sin^2\theta_c$
$\nu\bar{u} \to \mu^- \bar{d}$	$x\bar{u}(x)\cos^2\theta_c(1-y)^2$
$\nu\bar{u} \to \mu^- \bar{s}$	$x\bar{u}(x)\sin^2\theta_c(1-y)^2$
$\nu\bar{c} \to \mu^- \bar{s}$	$x\bar{c}(x)\cos^2\theta_c(1-y)^2$
$\nu\bar{c} \to \mu^- \bar{d}$	$x\bar{c}(x)\sin^2\theta_c(1-y)^2$
$\bar{\nu}u \to \mu^+ d$	$xu(x)\cos^2\theta_c(1-y)^2$
$\bar{\nu}u \to \mu^+ s$	$xu(x)\sin^2\theta_c(1-y)^2$
$\bar{\nu}c \to \mu^+ s$	$xc(x)\cos^2\theta_c(1-y)^2$
$\bar{\nu}c \to \mu^+ d$	$xc(x)\sin^2\theta_c(1-y)^2$
$\bar{\nu}\bar{d} \to \mu^+ \bar{u}$	$x\bar{d}(x)\cos^2\theta_c$
$\bar{\nu}\bar{s} \to \mu^+ \bar{u}$	$x\bar{s}(x)\sin^2\theta_c$
$\bar{\nu}\bar{s} \to \mu^+ \bar{c}$	$x\bar{s}(x)\cos^2\theta_c$
$\bar{\nu}\bar{d} \to \mu^+ \bar{c}$	$x\bar{d}(x)\sin^2\theta_c$

The QPM then expresses the cross-sections on the nucleon as the weighted sum of cross-sections on the quarks:

$$\frac{d^2\sigma}{dxdy} = \sum_i xq_i(x)\frac{d\sigma_i}{dy} = \sigma_0 \sum_i f_i(x, \cos\theta_c) \qquad (3.33)$$

where $\sigma_0 = G_F^2 s/2\pi(1 + Q^2/M_W^2)^2$ and the factors for each subprocess are listed in table 3.1.

It is then straightforward to write down the cross-sections for $\nu p, \nu n, \bar{\nu}p$ and $\bar{\nu}n$ so that for an isoscalar nucleon target

$$\frac{1}{\sigma_0}\frac{d^2\sigma^{\nu N}}{dxdy} = [xu(x) + xd(x) + 2xs(x)]$$
$$+ [x\bar{u}(x) + x\bar{d}(x) + 2x\bar{c}(x)](1-y)^2$$

$$\frac{1}{\sigma_0}\frac{d^2\sigma^{\bar{\nu} N}}{dxdy} = [x\bar{u}(x) + x\bar{d}(x) + 2x\bar{s}(x)]$$
$$+ [xu(x) + xd(x) + 2xc(x)](1-y)^2$$

$$(3.34)$$

Meanwhile we can write (2.41), for $s \to \infty$, in the form

$$\frac{d^2\sigma^{\nu N, \bar{\nu} N}}{dx dy} = \sigma_0 [\tfrac{1}{2}\{F_2(x) \pm xF_3(x)\} + \tfrac{1}{2}\{F_2(x) \mp xF_3(x)\}(1-y)^2]$$

(3.35)

Hence by taking $\bar{s}(x) = s(x), c(x) = \bar{c}(x)$ we get

$$F_2^{\nu N}(x) = F_2^{\bar{\nu} N} = x\Sigma(x)$$

(3.36)

and

$$\begin{aligned} xF_3^{\nu N}(x) &= xV(x) + 2[x\bar{s}(x) - x\bar{c}(x)] \\ xF_3^{\bar{\nu} N}(x) &= xV(x) - 2[x\bar{s}(x) - x\bar{c}(x)] \end{aligned}$$

(3.37)

i.e. in neutrino or antineutrino nucleon DIS, F_2 is the pure singlet combination and the average of xF_3 gives precisely the proton's valence quark distribution.

3.3.3 Neutrino-nucleon scattering via the neutral current

While the charged weak current is pure $V - A$ the neutral current is not, the hadronic neutral current being

$$J_\mu = \sum_i \tfrac{1}{2}\bar{q}(x)\gamma_\mu[(g_{Li} + g_{Ri}) - (g_{Li} - g_{Ri})\gamma_5]q(x)$$

(3.38)

where the g_{Li}, g_{Ri} are given above in terms of the Weinberg angle θ_W. So for elastic neutrino-quark scattering we pick up a constant part and a $(1-y)^2$ term:

$$\frac{d\sigma}{dy}(\nu q_i \longrightarrow \nu q_i) = \rho \frac{G_F^2 \hat{s}}{4\pi} [g_{Li}^2 + g_{Ri}^2(1-y)^2]$$

(3.39)

and ρ determines the relative strengths of the neutral and charged currents, being unity in the standard model. Using the QPM we can immediately write down the cross-sections for $\nu p, \nu n, \bar{\nu} p, \bar{\nu} n$ neutral current processes. As a result we get, for isoscalar nucleons:

$$F_2^{\nu N, \bar{\nu} N}(x) = x\Sigma(x) \sum_i \tfrac{1}{2}(g_{Li}^2 + g_{Ri}^2)$$

$$- 2[x\bar{s}(x) - x\bar{c}(x)]\tfrac{1}{4}(g_{Lu}^2 - g_{Ld}^2 + g_{Ru}^2 - g_{Rd}^2)$$

(3.40)

$$xF_3^{\nu N, \bar{\nu} N}(x) = xV(x) \sum_i \tfrac{1}{4}(g_{Li}^2 - g_{Ri}^2)$$

(3.41)

So F_2, xF_3, as in the charged current case, provide measures of the flavour singlet combination and valence distribution, respectively.

We can combine the neutral and charged current neutrino cross-sections to get, from (3.37) and (3.41), the Paschos-Wolfenstein (1973) relation

$$\frac{\sigma^{NC}(\nu N) - \sigma^{NC}(\bar{\nu} N)}{\sigma^{CC}(\nu N) - \sigma^{CC}(\bar{\nu} N)} = \sum_i \tfrac{1}{2}(g_{Li}^2 - g_{Ri}^2)$$

$$= \tfrac{1}{2} - \sin^2 \theta_W \tag{3.42}$$

thus allowing a clean determination of $\sin^2 \theta_W$.

3.3.4 Electron-nucleon scattering via the charged current

At HERA one can measure the processes $e^- p \to \nu X$ and $e^+ p \to \bar{\nu} X$ and even though there is no lepton detected in the final state, the values of Q^2, x and y can be obtained by measuring the jet of hadrons associated with the scattered quark. The formulas are given by (2.63). From the helicity point of view, $e^- q \to \nu q$ is just like $\nu q \to e^- q$ and $e^- \bar{q} \to \nu \bar{q}$ like $\nu \bar{q} \to e^- \bar{q}$. To get the relevant factors for each subprocess, we simply go to table 3.1 and reverse the direction of the process, at the same time changing the parton distribution to correspond to that of the initial parton. Thus for $e^- u \to \nu d$ we get a factor $x u(x) \cos^2 \theta_c$. The structure functions can then be immediately expressed in terms of the parton distributions,

$$F_2^{e^- p \to \nu X}(x) = x[u(x) + c(x) + \bar{d}(x) + \bar{s}(x)]$$
$$x F_3^{e^- p \to \nu X}(x) = x[u(x) + c(x) - \bar{d}(x) - \bar{s}(x)]$$
$$F_2^{e^+ p \to \bar{\nu} X}(x) = x[d(x) + s(x) + \bar{u}(x) + \bar{c}(x)]$$
$$x F_3^{e^+ p \to \bar{\nu} X}(x) = x[d(x) + s(x) - \bar{u}(x) - \bar{c}(x)] \tag{3.43}$$

3.4 Sum rules

Summing over all the contributing quarks must give the quantum numbers of the proton, in particular

$$\int_0^1 dx\, u_v(x) = 2 \ , \qquad \int_0^1 dx\, d_v(x) = 1 \tag{3.44}$$

The expressions for the structure functions in terms of the quark distributions, derived via the QPM, yield important sum rules.

Sometimes the origin of a sum rule is more fundamental than the QPM, e.g. the first one:

(a) Adler sum rule

$$\int_0^1 \frac{\mathrm{d}x}{x} (F_2^{\nu p}(x) - F_2^{\bar{\nu} p}(x)) = 2 \qquad (3.45)$$

This sum rule (Adler 1966) is for the charged current structure functions and follows from current conservation. It is exact and receives no QCD corrections.

(b) Bjorken sum rule

$$\int_0^1 \mathrm{d}x (F_1^{\bar{\nu} p}(x) - F_1^{\nu p}(x)) = 1 \qquad (3.46)$$

This sum rule (Bjorken 1967) does pick up a (negative) QCD correction – see section 5.3.

(c) Gross-Llewellyn Smith sum rule

$$\int_0^1 \mathrm{d}x (F_3^{\nu p}(x) + F_3^{\bar{\nu} p}(x)) = 6 \qquad (3.47)$$

This sum rule (Gross and Llewellyn Smith 1969) also gets a QCD correction.

(d) Gottfried sum rule

$$\int_0^1 \frac{\mathrm{d}x}{x} (F_2^{ep}(x) - F_2^{en}(x)) = \tfrac{1}{3} \qquad (3.48)$$

This sum rule (Gottfried 1967) requires the assumption of an $SU(2)$ symmetric sea, $\bar{d}(x) = \bar{u}(x)$.

(e) Momentum sum rule
For quarks of type a, the fraction of the proton's momentum carried is just $\int \mathrm{d}x x q_a(x)$. The sum of these must be less than unity,

$$\int_0^1 \mathrm{d}x x \Sigma(x) < 1 \qquad (3.49)$$

From experiment, this fraction seems close to $\tfrac{1}{2}$. The remaining momentum is carried by the gluons. In fact QCD fixes the asymptotic value of this fraction to be $(1 + 16/3f)^{-1}$, where f is the number of quark flavours.

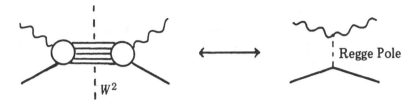

Fig. 3.4   Forward amplitude for virtual photon nucleon scattering, as $W^2 \to \infty$.

3.5   Behaviour of quark distributions at small and large x

3.5.1   Small x

For large Q^2, the limit $x \to 0$ corresponds to the high energy (large W^2) region for the virtual photon nucleon cross-section. From our experience with the asymptotic behaviour of total cross-sections, we expect the W^2 dependence to be governed by the exchange of the leading Regge pole in the t-channel of the elastic amplitude – see fig. 3.4.

In the W^2 or $\nu \to \infty$ limit we have

$$\begin{aligned}
W_1(\nu, Q^2) &\sim (W^2)^\alpha f_\alpha^1(Q^2) = \nu^\alpha f_\alpha^1(Q^2) \\
W_2(\nu, Q^2) &\sim (W^2)^{\alpha-2} f_\alpha^2(Q^2) = \nu^{\alpha-2} f_\alpha^2(Q^2) \\
W_3(\nu, Q^2) &\sim (W^2)^{\alpha-2} f_\alpha^3(Q^2) = \nu^{\alpha-2} f_\alpha^3(Q^2)
\end{aligned} \qquad (3.50)$$

where α is the intercept of the leading Regge exchange; $\alpha = 1$ for the Pomeron, $\alpha \simeq \frac{1}{2}$ for the $\rho - f - w - A_2$ exchange. In the QPM, the structure functions scale and this determines the large Q^2 behaviour of the f_α^i. That is

$$f_\alpha^1(Q^2) \sim (Q^2)^{-\alpha} \ , \quad f_\alpha^2(Q^2), \ f_\alpha^3(Q^2) \sim (Q^2)^{1-\alpha}$$

and so

$$F_1(x) \sim x^{-\alpha} \ , \quad F_2(x) \sim x^{1-\alpha} \qquad (3.51)$$

For the flavour singlet combination we have $I = 0$ exchange in the t-channel and the leading exchange is the Pomeron. For non-singlet combinations, flavour is exchanged and the leading trajectory has $\alpha = \frac{1}{2}$. So we expect

$$x\Sigma(x) \sim \text{const}, \qquad xV(x) \sim x^{\frac{1}{2}} \text{ as } x \to 0 \qquad (3.52)$$

This means that near $x = 0$, the dominant contribution to the structure function is from the sea. Note that (3.52) implies an infinite number of sea quarks in the nucleon since $\int \mathrm{d}x S(x)$ will diverge logarithmically.

3.5.2   Large x

We saw in section 2.2 that $x = 1$ corresponds to $W^2 = M^2$, i.e. elastic scattering and that the large Q^2 behaviour of the structure function is determined by the nucleon form factor

$$W_1(\nu, Q^2) = \delta(W^2 - M^2)\frac{Q^2}{2M}G_M^2(Q^2) \qquad (3.53)$$

Fig. 2.2 shows that, at low values of Q^2, the structure function is built up from resonance contributions as $x \longrightarrow 1$, i.e. the form factors of nucleon resonances determine the Q^2 behaviour. But as Q^2 increases, these contributions fall rapidly, displaying a smooth behaviour. If scaling should happen to hold at these small values of Q^2, it means that the contributions from the resonances should, on average, describe the large x behaviour. Empirically it was found that this duality was best seen in a variable x' (Bloom and Gilman 1971) rather than x, where $x' = x[1 + m_0^2 x/Q^2]^{-1}$ where $m_0 \simeq 1.1$ GeV. This correspondence between the two descriptions of the large x region can be written

$$\int \mathrm{d}x' F_1(x') \propto \sum_{res} \frac{Q^2}{2M}G_{M,res}^2(Q^2) \qquad (3.54)$$

Suppose we can write $F_1(x) \sim (1 - x')^n$ and $G_M(Q^2) \sim (Q^2)^{-N}$ then since $(1 - x) \sim 1/Q^2$ we get the relation

$$n = 2N - 1 \qquad (3.55)$$

known as the Drell-Yan-West relation (Drell and Yan 1970, West 1970). The dipole form factor implies $N = 2$ so that (3.55) gives

$$F_1(x), \quad F_2(x) \sim (1 - x)^3 \text{ as } x \longrightarrow 1 \qquad (3.56)$$

Empirical relations for the exponents governing (a) the Q^2 dependence of hadron form factors, (b) the large x behaviour of structure functions and (c) the energy dependence of fixed angle scattering of hadrons are known as the constituent counting rules (Brodsky and Farrar 1973, Matveev, Murddyan and Tavkheldize

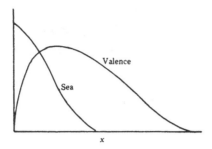

Fig. 3.5   Qualitative behaviour expected from the two components of quark distributions based on the discussion of section 3.5.

1973). The rules say that the relevant exponent in each case depends simply on the minimum number of elementary constituents in the hadronic bound system. They are based on the assumption that the particular bound state is a collinear system of elementary constituents, one or more of which undergoes hard scattering. Clearly QCD is the framework in which to explore such ideas, and we return to a discussion of the rules in chapter 6.

For the moment we simply state the results for the cases of interest here. The large Q^2 behaviour of the form factor for elastic scattering of a hadron made of a minimum number n_H of elementary constituents is

$$F_H(Q^2) \sim (Q^2)^{1-n_H} \qquad (3.57)$$

The $x \to 1$ behaviour of the hadronic structure function is governed by the number n_s of 'spectator' constituents of the hadron,

$$F_2(x) \sim (1-x)^{2n_s - 1} \qquad (3.58)$$

For a hadron with n_H quarks, the diagram of fig. 3.1 gives $n_s = n_H - 1$ which makes (3.57), (3.58) consistent with (3.56). Relations (3.57), (3.58) are the simplest versions of the constituent counting rules. The rules can be extended to the large x behaviour of the sea quarks, $x\bar{q}(x) \sim (1-x)^{\bar{n}}$ with $\bar{n} = 5 - 7$.

As a result of the discussion of the $x \to 0, 1$ behaviour we arrive at a qualitative picture for the two components of the structure functions shown in fig. 3.5.

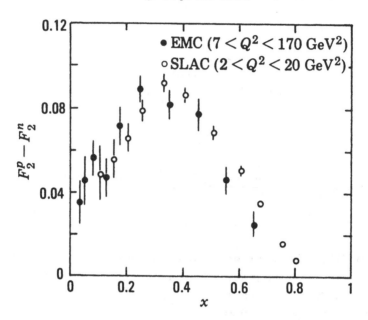

Fig. 3.6 Data from the EMC (Aubert *et al.* 1987) on $F_2^{\mu p} - F_2^{\mu n} \sim x(u_v(x) - d_v(x))$ averaged over Q^2, compared with data from SLAC at lower Q^2.

3.6 Information on quark distributions from experiment

3.6.1 *Information from electromagnetic structure functions*

The difference of the proton and neutron structure functions given by (3.21) allows a determination of a valence combination $x(u_v(x) - d_v(x))$. Fig. 3.6 shows data on this quantity obtained from muons scattered off deuterium and hydrogen. The data certainly are consistent with an $x^{\frac{1}{2}}$ behaviour as $x \to 0$.

Going to large x, the relative magnitudes of the u and d distributions can be estimated from the ratio of the neutron and proton structure functions. For, putting $S(x) = 0$ gives

$$\frac{F_2^{\mu n}(x)}{F_2^{\mu p}(x)} = \frac{1 + 4d_v(x)/u_v(x)}{4 + d_v(x)/u_v(x)} \tag{3.59}$$

and if $SU(6)$ were exact, $d_v(x)/u_v(x) = \frac{1}{2}$ and then $F_2^n(x)/F_2^p(x) \to \frac{2}{3}$. The data shown in fig. 3.7 however are consistent with $n/p \to \frac{1}{4}$ as $x \to 1$ linearly, i.e. consistent with $d_v(x)/u_v(x) \sim (1-x)^1$ as $x \to 1$.

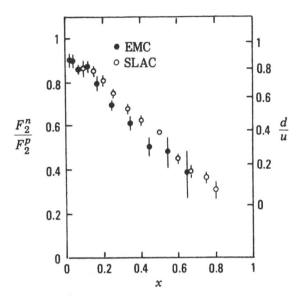

Fig. 3.7 Data from the EMC collaboration (Aubert *et al.* 1987) on $F_2^{\mu n}/F_2^{\mu p}$ together with SLAC data on F_2^{en}/F_2^{ep}.

There is a great amount of data on electromagnetic structure functions from muon experiments at CERN. For a comprehensive survey, see the review by Sloan, Smadja and Voss (1988).

3.6.2 Information from weak structure functions in neutrino scattering

From (3.34) we see that it is possible to extract a pure antiquark distribution from antineutrino cross-sections, e.g. by selecting events where y is close to 1. Then $\sigma^{\bar{\nu}N}$ picks out the combination $\bar{u} + \bar{d} + 2\bar{s}$. If we take a combination of $\sigma^{\bar{\nu}N}$ and $\sigma^{\nu N}$ given by

$$\frac{1}{\sigma_0}\left[\frac{\mathrm{d}^2\sigma^{\bar{\nu}N}}{\mathrm{d}x\mathrm{d}y} - (1-y)^2\frac{\mathrm{d}^2\sigma^{\nu N}}{\mathrm{d}x\mathrm{d}y}\right]$$

$$= x[\bar{u}(x) + \bar{d}(x) + 2\bar{s}(x)][1 - (1-y)^4]$$
$$+ 2x[\bar{c}(x) - \bar{s}(x)](1-y)^2[1 - (1-y)^2] \qquad (3.60)$$

then we get information on the non-charm sea by (a) assuming we are below threshold for exciting charm quarks, $\bar{c}(x) = 0$ or (b) combining information on $x\bar{s}(x)$ from dimuon production. The extra muon is attributed to the decay of a produced charm quark,

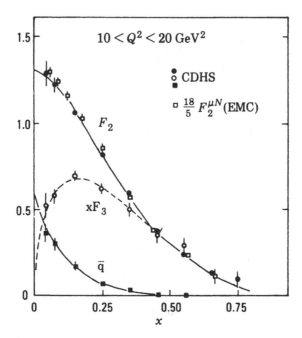

Fig. 3.8 Data from the CDHS collaboration on neutrino structure functions together with $F_2^{\mu N}$. The neutrino data are from Abramowicz *et al.* (1983), using a target of iron.

i.e. $\bar{\nu}s \rightarrow \mu^+\bar{c}$ followed by $\bar{c} \rightarrow \bar{s}\mu^-\bar{\nu}$. To extract the strange quark distribution we need to take into account the mass of the charm quark and the threshold effect it causes.

In this situation, the fractional momentum of the struck quark (s) is no longer x but ξ where $\xi = x + m_c^2/2M\nu = x(1+m_c^2/Q^2)$, m_c being the charm quark mass. The 'slow rescaling' prescription (Brock 1980) uses the variable ξ to modify the QPM (a) by replacing $xs(x)$ by $\xi s(\xi)$ and (b) expressing the Callan-Gross relation (3.15) in terms of ξ. From (b), the quark distribution gets multiplied by $(1-y+xy/\xi)$. The strange quark distribution obtained in this way could be softer than the non-strange sea and is probably suppressed; fits to the dimuon data suggest $2\bar{s}/(\bar{u}+\bar{d}) \simeq \frac{1}{2}$.

Some structure functions obtained from νN, $\bar{\nu}N$ data are shown in fig. 3.8, including the antiquark combination $x(\bar{u}+\bar{d}+2\bar{s})$. From (3.22) and (3.36) we get a relation between the weak and electromagnetic structure functions

$$F_2^{\nu N}(x) = \tfrac{18}{5}F_2^{\mu N} + \tfrac{6}{5}[x\bar{s}(x) - x\bar{c}(x)] \qquad (3.61)$$

The figure shows the l.h.s. of (3.61) and the first term on the r.h.s; the near equality of the two providing beautiful confirmation of the predictions of the QPM.

3.7 Polarised structure functions

Let us consider the QPM predictions for the polarised structure functions G_1 and G_2 defined in section (2.4). The QPM expresses the antisymmetric part of the hadronic tensor as

$$W_{\mu\nu}^{(A)}(q,p,S) = \sum_i \sum_s \int \mathrm{d}^4k \; f_s^i(p,k,S) \; w_{\mu\nu,s}^{(A)i}(q,k) \; \delta[(k+q)^2] \qquad (3.62)$$

where S is the polarisation vector of the proton and the summations are over the quark flavours and helicities, as before. As in section 3.2 we consider massless on-shell quarks. The f_s^i here are the same distributions as in section 3.2 but the S dependence was then omitted since we summed over the proton polarisations. The $w_{\mu\nu,s}^{(A)i}$ tensor is the antisymmetric part of the photon-quark interaction given by

$$\begin{aligned} w_{\mu\nu,s}^{(A)i}(q,k) &= \frac{1}{4}e_i^2 \; \mathrm{Tr}[(1 - s\gamma_5) \; k\!\!\!/\gamma_\mu(k\!\!\!/ + q\!\!\!/)\gamma_\nu] \\ &= \mathrm{i}e_i^2 \; s \; \epsilon_{\mu\nu\rho\sigma}q_\rho k_\sigma \end{aligned} \qquad (3.63)$$

In the covariant QPM, f_s^i is a function only of Lorentz invariants and since $p \cdot S = 0$ we have only $p \cdot k$, $k \cdot S$, $p^2 = M^2$. So we can write

$$w_{\mu\nu,s}^{(A)}(q,p,S) = \mathrm{i}\epsilon_{\mu\nu\rho\sigma}q_\rho \sum_i e_i^2 \int \mathrm{d}^4k \; \Delta f^i(p \cdot k, k \cdot S) \; k_\sigma \; \delta[(k+q)^2] \qquad (3.64)$$

where $\Delta f^i = f_+^i - f_-^i$. Following Jackson, Ross and Roberts (1989) we observe that Δf^i must be proportional to $k \cdot S$. This can be seen from realising that p, k, q are odd under parity and S is even; to make $W_{\mu\nu}^{(A)}$ odd (as demanded by 2.52) and, since the proton has spin $\tfrac{1}{2}$, the dependence on S must be simply $(k \cdot S)$.

Thus we write

$$\Delta f^i(p \cdot k, k \cdot S) = -\frac{(k \cdot S)}{M} \tilde{f}_i(p \cdot k) \qquad (3.65)$$

Since $k \cdot S$ just projects the helicity of the parton along the direction of the proton's polarisation then we must have $k_T \neq 0$ in order to make this projection along any given direction. Again we choose x_i to be the fraction of the proton light-cone momentum carried by the quark, $x_i = k^+/p^+$, $k^- = k_T^2/2k^+$, and $2p \cdot k/M^2 = x + k_T^2/xM^2$.

Let us consider first the case of a longitudinally polarised proton where $S = (0, 0, 0, 1)$ in the proton rest frame. We have $k \cdot S = -\frac{1}{2}xM(1 - (k_T/x^2 M^2))$. Taking $\mu = 1, \nu = 2$ in (2.52) gives $W_{12}^{(A)} = -i\frac{\nu}{M^2}G_1(\nu, Q^2)$. Also $\epsilon_{12\rho\sigma}q_\rho k_\sigma$ in (3.64) gives $q_3 k_0 - q_0 k_3 \rightarrow -x\nu M$. Thus we get

$$\frac{\nu}{M}G_1(\nu, Q^2) = \frac{1}{4}\sum_i e_i^2 \int d^4k \; x^2 \; (1 - \frac{k_T^2}{x^2 M^2})$$

$$\times \tilde{f}_i(x + \frac{k_T^2}{xM^2}) \; \delta(x_i - x)$$

$$= \frac{\pi x}{8}\sum_i e_i^2 \int dk_T^2 \; (1 - \frac{k_T^2}{x^2 M^2}) \; \tilde{f}_i(x + \frac{k_T^2}{xM^2})$$

$$\qquad (3.66)$$

The r.h.s. of (3.66) depends on x rather than ν, Q^2 separately and so we get the scaling result

$$\frac{\nu}{M}G_1(\nu, Q^2) \longrightarrow g_1(x) \qquad (3.67)$$

Now consider the case of a transversely polarised proton where $S = (0, 1, 0, 0)$ and $k \cdot S = -k_x$. Taking $\mu = 0, \nu = 2$ in (2.52) gives $W_{02}^{(A)} = -i\{\frac{\nu}{M^2}G_1 + \frac{\nu^2}{M^3}G_2\}$. We also get $\epsilon_{02\rho\sigma}q_\rho k_\sigma = \nu k_x$ and using $k_x^2 = \frac{1}{2}k_T^2$ we end up with

$$\frac{\nu}{M}G_1(\nu, Q^2) + \frac{\nu^2}{M^2}G_2(\nu, Q^2)$$

$$= \frac{x}{8}\sum_i e_i^2 \int dk_T^2 \; (\frac{k_T^2}{x^2 M^2}) \; \tilde{f}_i(x + \frac{k_T^2}{xM^2}) \quad (3.68)$$

giving the scaling result

$$\frac{\nu^2}{M^2}G_2(\nu, Q^2) \longrightarrow g_2(x) \qquad (3.69)$$

Thus we see the explicit dependence on parton transverse momentum, k_T, in the spin structure function $g_1(x)$ and $g_2(x)$. If we differentiate (3.68) w.r.t. x we obtain (3.66) divided by $-x$; i.e. we have

$$-g_1(x) = x\frac{\partial}{\partial x}\{g_1(x) + g_2(x)\} \qquad (3.70)$$

or

$$g_2(x) = \int_x^1 \frac{dy}{y} g_1(y) - g_1(x) \qquad (3.71)$$

From (3.71) follows a set of sum rules (Wandzura and Wilczek 1977),

$$\int_0^1 dx\, x^{J-1} \{\tfrac{J-1}{J} g_1(x) + g_2(x)\} = 0 \qquad (3.72)$$

where $J = 1$ gives the Burkhardt-Cottingham (1970) sum rule

$$\int_0^1 dx g_2(x) = 0 \qquad (3.73)$$

Thus we see that the longitudinal polarisation of the proton is described by $g_1(x)$ and the transverse polarisation by $g_1(x)+g_2(x)$. Since $g_2(x)$ can be computed from $g_1(x)$ via (3.71), it follows that the measurement of the longitudinal polarisation also determines $g_2(x)$ and, in turn, the tranverse polarisation. In fig. 3.9 the values of $g_1(x)$ obtained from the measurement of the longitudinal polarisation (Ashman *et al.* 1988a,1989) together with the resulting estimates for $g_2(x)$ and the transverse polarisation.

In contrast to the relation between F_1 and F_2, the relation (3.71) between g_1 and g_2 relies on the assumption of massless on-shell partons. When we allow $k^2 \neq 0$, the k_σ appearing in (3.63), which represents the polarisation vector, becomes $\sim k_\sigma + (k^2/p\cdot k)p_\sigma$ for small k^2. As a result, correction terms like $k^2/x^2 M^2$ arise, spoiling the precise connection between g_1 and g_2.

Next we consider sum rules involving g_1. Take a proton with component of spin $S_z = \tfrac{1}{2}$ along the z-axis. The current $J_\mu^3 \sim 2S_z$ is related, through an isospin rotation, to the current J_μ^+ which enters in nucleon β-decay; $< p|\bar{u}\gamma_\mu(1 - \gamma_5)d|n > = \bar{u}_p\gamma_\mu(1 - (g_A/g_V)\gamma_5)u_n$. From this we have

$$\left|\frac{g_A}{g_V}\right|_{np} = 2S_z(u_+ - d_+) \qquad (3.74)$$

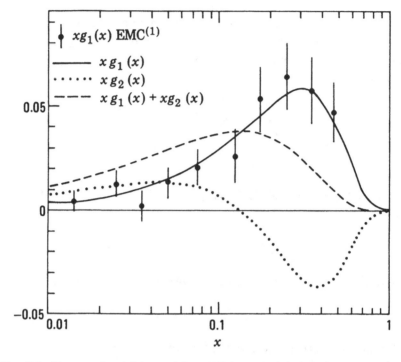

Fig. 3.9 Expectations for $xg_2(x)$, $xg_1(x) + xg_2(x)$ based on taking a parametrisation of the asymmetry $A(x) = g_1(x)/F_1(x)$ together with the measured values of $g_1(x)$.

where $q_+ = q + \bar{q}$ and $S_z(q)$ means the z component of the quark spin. Similarly one can relate the axial-vector couplings for hyperon decays to other quark combination spins, e.g.

$$3\left|\frac{g_A}{g_V}\right|_{\Lambda p} = 2S_z(2u_+ - d_+ - s_+)$$

$$3\left|\frac{g_A}{g_V}\right|_{\Xi\Lambda} = 2S_z(u_+ + d_+ - 2s_+) \qquad (3.75)$$

$$\left|\frac{g_A}{g_V}\right|_{\Sigma n} = 2S_z(d_+ - s_+)$$

In Cabibbo theory there are only two independent combinations, reflecting the fact that there are just two independent couplings F, D for the octet of weak currents. Thus the three quantities on the l.h.s. of (3.69) are respectively $3F + D$, $3F - D$ and $F - D$, while $|g_A/g_V|_{np}$ is $F+D$. Now returning to the polarised structure function g_1, we can write for the proton

$$\int_0^1 \mathrm{d}x \, g_1^p(x) = \tfrac{1}{2} \int_0^1 \mathrm{d}x [\tfrac{4}{9}\Delta u_+(x) + \tfrac{1}{9}\Delta d_+(x) + \tfrac{1}{9}\Delta s_+(x)] \quad (3.76)$$

assuming three flavours only. Here $\Delta q = q \uparrow -q \downarrow$ where $\uparrow, \downarrow$ refer to the component of spin parallel or antiparallel to the proton spin. The expression for the neutron is obtained by interchanging $u \leftrightarrow d$ in (3.76) and if we use $2S_z(q) = \int \mathrm{d}x \, q(x)$ we get

$$\begin{aligned}
\int_0^1 \mathrm{d}x \, g_1^{p,n}(x) &= \pm \tfrac{1}{12}[2S_z(u_+ - d_+)] + \tfrac{1}{36}[2S_z(u_+ + d_+ - 2s_+)] \\
&\quad + \tfrac{1}{9}[2S_z(u_+ + d_+ + s_+)] \\
&= \pm T_3 + T_8 + T_0
\end{aligned}$$

$$(3.77)$$

where T_3, T_8, T_0 are the flavour triplet, octet and singlet combinations. So taking the difference of proton and neutron gives the Bjorken (1966, 1970) sum rule

$$\begin{aligned}
\int_0^1 \mathrm{d}x [g_1^p(x) - g_1^n(x)] = 2T_3 &= \tfrac{1}{6} [2S_z(u_+ - d_+)] \\
&= \frac{1}{6} \left| \frac{g_A}{g_V} \right|_{np}
\end{aligned}$$

$$(3.78)$$

There are QCD corrections (see section 5.3) to the r.h.s. of (3.78). Experimentally it is difficult to set up a polarised target to obtain a measurement of g_1^n so Ellis and Jaffe (1974) derived a sum rule for g_1^p alone. If we *assume* that the strange quark sea is unpolarised then we have in (3.77)

$$T_8 = \frac{1}{36} \left| \frac{g_A}{g_V} \right|_{np} \cdot \frac{3F - D}{F + D} \;, \quad T_0 = \frac{1}{9} \left| \frac{g_A}{g_V} \right|_{np} \cdot \frac{3F - D}{F + D} \quad (3.79)$$

giving

$$\int_0^1 \mathrm{d}x \, g_1^p(x) = \frac{1}{12} \left| \frac{g_A}{g_V} \right|_{np} \left[1 + \frac{5}{3} \cdot \frac{3F - D}{F + D} \right] \quad (3.80)$$

Clearly the assumption $S_z(s_+) = 0$ is a strong one but it raises the question whether a measurement of the l.h.s. of (3.80) can give information of the spin content of the proton. Let us define $\Delta\Sigma = 2S_z(u_+ + d_+ + s_+)$ which corresponds to the fraction of the spin of proton carried by the quarks. From (3.75), (3.77) we can write, for example

$$\int_0^1 dx\, g_1^p(x) = \frac{1}{18}\left[\left|\frac{g_A}{g_V}\right|_{\Sigma n} + 2\left|\frac{g_A}{g_V}\right|_{np}\right] + \frac{1}{9}\,\Delta\Sigma \qquad (3.81)$$

or in terms of the other measured g_A/g_V in hyperon decays. Actually the QCD corrections to (3.80), (3.81) are more complicated since the flavour singlet, T_0, receives a correction radically different to the non-singlet pieces T_3, T_8.

In fig. 3.9 the EMC values of $g_1(x)$ are shown and so the evaluation of the integral on the l.h.s. of (3.81) is thus the area under the data shown in the figure. Assuming an extrapolation to $x = 0$ given by $xg_1(x) \sim x$ yields a value of the integral $\int d(\ln x)xg_1(x)$ very close to the experimentally measured value of the first term on the r.h.s. of (3.81). This implies that $\Delta\Sigma$ must be close to zero, i.e. a considerable part of the proton's spin would have to be carried by gluons and/or by orbital angular momentum.

3.8   Parton transverse momentum

From the discussion in the previous section, it is clear that parton transverse momentum, k_T, plays a crucial role for the structure functions, g_1, g_2, of the polarised proton. The measurement of g_1 may therefore be expected to provide information on the average value of k_T for example. We shall see that this is indeed the case and, furthermore, estimates of $< k_T^2 >$ can be extracted from the *un*polarised structure functions.

First consider the helicity-weighted average of k_T given by (Jackson, Ross and Roberts 1989)

$$< \mathbf{k}_T \cdot \mathbf{S}_T >_{hw} = \frac{-\int dk_T^2\, (k \cdot S_T)\, [f_+(p \cdot k) - f_-(p \cdot k)]}{\int dk_T^2\, [f_+(p \cdot k) + f_-(p \cdot k)]} \qquad (3.82)$$

where $\mathbf{S}_T$ is a unit vector in the spin direction of a transversely polarised proton. The numerator of the r.h.s. of (3.82) is proportional to the r.h.s. of (3.68) while the denominator is proportional the r.h.s. of (3.10). Thus we can write

$$< \mathbf{k}_T \cdot \mathbf{S}_T >_{hw} = \frac{2x^2 M[g_1(x) + g_2(x)]}{F_2(x)} \qquad (3.83)$$

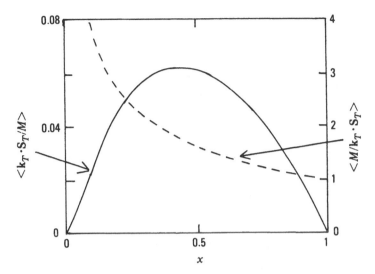

Fig. 3.10   Helicity-weighted averages of (3.83),(3.84) as a function of x. Taken from Jackson, Ross and Roberts 1989.

A helicity-weighted average of k_T^{-1} can be defined likewise; adding (3.66) and (3.68) gives

$$< (\mathbf{k}_T \cdot \mathbf{S}_T)^{-1} >_{hw} \; = \; \frac{2[2g_1(x) + g_2(x)]}{MF_2(x)} \qquad (3.84)$$

In fig. 3.10 these two quantities are estimated using parametrisations of the EMC data.

While the average of k_T^{-1} of (3.84) turns out to depend on the assumption of massless on-shell partons, one can argue that there is no evidence for significant corrections to this estimate. In this case, (3.83) provides a *lower* bound on $< k_T >$ while (3.84) provides an *upper* bound on $< (k_T)^{-1} >^{-1}$.

While the role of k_T is so prominent in the polarised structure functions, it is apparently less obvious in $F_1(x)$ or $F_2(x)$. However, we can rewrite (3.9),(3.10) in the form

$$F_1(x) \; = \; \frac{\pi M^2 x}{4} \int_x^1 \mathrm{d}y \; [f_+(y) + f_-(y)] \qquad (3.85)$$

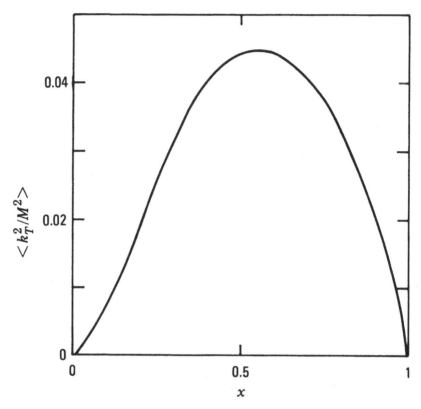

Fig. 3.11   Mean square transverse momentum of partons as a function of x given by (3.87), taken from Jackson, Ross and Roberts (1989).

where $y = 2p \cdot k/M^2 = x + k_T^2/xM^2$. Defining the average of k_T^2 by

$$< k_T^2 > = \frac{\int \mathrm{d}k_T^2 \, k_T^2 \, [f_+(y) + f_-(y)]}{\int \mathrm{d}k_T^2 \, [f_+(y) + f_-(y)]} \qquad (3.86)$$

and using $F_2(x) = 2xF_1(x)$ we get

$$< k_T^2 > = \frac{x^3 M^2}{F_2(x)} \int_x^1 \mathrm{d}y \, \frac{F_2(y)}{y^2} \qquad (3.87)$$

and the result is shown in fig. 3.11.

This result is phenomenologically similar to some other estimates (Close, Halzen and Scott 1977). Landshoff (1977) obtained an estimate for $< k_T^2 >$ also by using the covariant parton model but by making assumptions about the recoiling 'core' (X

of fig. 3.1) rather than about the struck parton. The result shown in fig. 3.11 is quite consistent with the bounds shown in fig. 3.10 on k_T obtained from the polarised scattering measurements. In fact, one finds that, to a good approximation,

$$< \mathbf{k}_T \cdot \mathbf{S}_T >_{hw} < (\mathbf{k}_T \cdot \mathbf{S}_T)^{-1} >_{hw} \simeq < k_T^2 > \qquad (3.88)$$

An independent estimate of the average transverse momentum of partons in DIS comes from studying the azimuthal dependence of the scattered quark. Cahn (1977) retained terms which $\sim 1/Q$ in $l^{\mu\nu} W_{\mu\nu}$ and showed that

$$< \cos \phi > = -(\frac{2p_T}{Q}) \frac{(2-y)\sqrt{1-y}}{1+(1-y)^2} \qquad (3.89)$$

where ϕ is the angle between the lepton scattering plane and the plane of the scattered quark. Measurements of the azimuthal dependence of the hadrons associated with quark-jet yield estimates for $< k_T >$, via (3.89), which are 200–400 MeV (Renton 1986).

4

Perturbative QCD

In the impulse approximation of the QPM the quarks in the proton behave as though very weakly bound. From this followed the scaling behaviour of the structure functions. We now introduce the underlying field theory for describing quark interactions, QCD, which reproduces the QPM at asymptotic energies. This feature of describing the quarks as free particles at such energies is the famous 'asymptotic freedom' property of non-Abelian theories such as QCD. In this chapter we introduce the basics of perturbative QCD necessary to describe DIS.

4.1 Renormalisation group and the running coupling constant

QCD is a Yang–Mills theory where the gauged symmetry is colour. The Lagrangian is

$$\mathcal{L} = -\tfrac{1}{2}\, G^a_{\mu\nu}\, G^{\mu\nu}_a \; + \; \sum_k^{N_f} \bar{q}_k (\mathrm{i}\gamma^\mu D_\mu - m_k) q_k \qquad (4.1)$$

where q_k is the quark field, $k = 1, \ldots, N_f$ and a, the colour index, $= 1, \ldots, 8$ and we shall assume zero mass quarks, $m_k = 0$. Here

$$G^a_{\mu\nu} = \partial_\mu A^a_\nu - \partial_\nu A^a_\mu + g\, f_{abc}\, A^b_\mu A^c_\nu \qquad (4.2)$$

and

$$D_\mu \equiv \partial_\mu - \mathrm{i} g\, T^a A^a_\mu \qquad (4.3)$$

where A^a_μ is the μth component of the ath gluon vector field. The matrices T^a represent the gluons and form an $SU(3)$ algebra,

$$[T^a, T^b] = \mathrm{i} f_{abc}\, T^c \qquad (4.4)$$

50

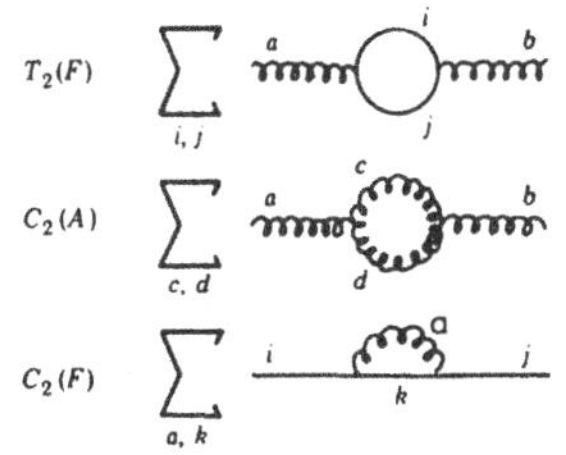

Fig. 4.1 Colour factors associated with loops, $C_2(F) = \frac{4}{3}$, $C_2(A) = 3$ and $T_2(F) = \frac{1}{2}$ in $SU(3)$ of colour.

and the convention is to take them proportional to the Gell-Mann matrices, $T^a = \frac{1}{2}\lambda^a$ so that

$$\mathrm{Tr}(T^a T^b) = T_2(F)\,\delta_{ab} = \tfrac{1}{2}\delta_{ab} \tag{4.5}$$

The factor $T_2(F)$ normalises the bare coupling g. We shall also need the Casimirs $C_2(F)$ and $C_2(A)$ given by

$$\sum_a^8 \sum_k^N T^a_{ik} T^a_{kj} = C_2(F)\delta_{ij} = \tfrac{4}{3}\delta_{ij} \tag{4.6}$$

$$\sum_{c,d}^8 f_{acd} f_{bcd} = C_2(A)\delta_{ab} = 3\delta_{ab} \tag{4.7}$$

The factors $T_2(F)$, $C_2(F)$ and $C_2(A)$ are just the colour factors associated with the loops shown in fig. 4.1

In the bare Lagrangian of (4.1) for massless quarks (the gluons are massless to maintain gauge invariance) there is no scale, the bare coupling g being dimensionless. If we take the quark–gluon coupling with external momenta $\sim Q$ and compute the loop corrections (expressed in terms of a large momentum cut-off Λ) we get an effective coupling $\bar{g}$ of the form

$$\bar{g} \equiv \bar{g}(Q^2)$$
$$= g - \beta_0\,\frac{g^3}{32\pi^2}\left[\ln\frac{Q^2}{\Lambda^2} + \kappa\right] + \tfrac{3}{2}\,\beta_0^2\,\frac{g^5}{(32\pi^2)^2}\,\ln^2\frac{Q^2}{\Lambda^2} + \cdots \tag{4.8}$$

This is one example of a bare Green's function all of which diverge in the theory. Here, $\bar{g}(Q^2) \to \infty$ as $\Lambda \to \infty$. However if we could somehow 'fix' the coupling to be finite for some choice $Q^2 = \mu^2$ then (4.8) would allow us to express the coupling at *any* value of

Q^2 in terms of this finite coupling,

$$\bar{g}(Q^2) = \bar{g}(\mu^2) - \frac{\beta_0}{32\pi^2}\,\bar{g}(\mu^2)\left[\ln\frac{Q^2}{\mu^2} + \kappa\right] + \ldots \qquad (4.9)$$

What has been gained is that the effective coupling $\bar{g}(Q^2)$ is now manifestly independent of the unphysical quantity Λ but the price paid is the introduction of a scale μ^2 at which the effective coupling is renormalised.

A formal way to proceed with the renormalisation of the Green's functions in a theory is via the renormalisation group equation (RGE). To begin with, consider a theory with only one bare scalar field, ϕ_0, the self-interaction characterised by a bare coupling g_0. The *renormalised* field ϕ is defined via a scale factor Z_ϕ.

$$\phi_0 = Z_\phi(g_0, \frac{\Lambda}{\mu})\,\phi \qquad (4.10)$$

where Λ, μ have the same meaning as above. From dimensional arguments, Z_ϕ depends only on their ratio. The renormalised and bare Lagrangians, $\mathcal{L}, \mathcal{L}_0$ describe the same physical quantities and, as a result, we can relate renormalised and bare Green's functions by

$$\Gamma^{(n)}(p, g, \mu) = Z_\phi^n(g_0, \frac{\Lambda}{\mu})\,\Gamma_0^{(n)}(p, g_0, \Lambda) \qquad (4.11)$$

That is, the renormalised Green's function (with n external fields and external momentum p) is independent of the u.v. cut-off Λ but needs a momentum scale μ where one chooses to normalise $\Gamma^{(n)}$. Now consider (4.11) under a change of this scale $\mu \to \mu + d\mu$. The constraint $\mu d\Gamma^{(n)}/d\mu = 0$ gives

$$\left[\mu\frac{\partial}{\partial\mu} + \beta(g)\frac{\partial}{\partial g} - n\gamma_\phi(g)\right]\Gamma^{(n)}(p, g, \mu) = 0 \qquad (4.12)$$

where

$$\gamma_\phi(g) \equiv \mu\frac{\partial}{\partial\mu}\left[\ln Z_\phi(g_0, \frac{\Lambda}{\mu})\right]\Bigg|_{g_0,\Lambda} \qquad (4.13)$$

and

$$\beta(g) \equiv \mu\frac{\partial}{\partial\mu}g(\mu)\Bigg|_{g_0,\Lambda} \qquad (4.14)$$

Equation (4.12) simply expresses the way that g, ϕ must vary to compensate for the given change to μ. This is the renormalisation group equation (RGE) (Stückelberg and Peterman 1953, Gell-Mann and Low 1954, Callan 1970, Symanzik 1970). In general there is one β-function for each coupling and one *anomalous dimension*, $\gamma(g)$ for each field. Thus in QCD we have $\gamma_F(g)$ and $\gamma_A(g)$ associated with the quark and gluon fields and one β-function. The β-function is important in determining the asymptotic behaviour of the effective coupling and the anomalous dimensions determine the precise high energy behaviour of Green's functions. To study the solutions of the RGE, first put $\gamma_\phi(g) = 0$ in (4.12). Changing the external momentum $p \to \sigma p = e^t p$ implies that the change to μ must be $\mu \to \mu e^{-t}$ (from dimensions) and (4.12) demands that this change is compensated for by a change in the coupling $g \to \bar{g}(t)$ where $\bar{g}$ is the solution of

$$t = \int_g^{\bar{g}(t)} \frac{dg}{\beta(g)} \qquad (4.15)$$

or

$$\frac{d\bar{g}(t)}{dt} = \beta(t) \text{ with } \bar{g}(0) = g \qquad (4.16)$$

In this case the solution of the RGE is simply

$$\Gamma(e^t p, g, \mu) = \Gamma(p, \bar{g}(t), \mu) \qquad (4.17)$$

The change of momentum scale can thus be absorbed by a description in terms of an *effective* coupling $\bar{g}$ which 'runs' with the value of the scale change. For $\gamma_\phi(g) \neq 0$ the solution becomes

$$\Gamma^{(n)}(e^t p, g, \mu) = \Gamma^{(n)}(p, \bar{g}(t), \mu) \exp\left\{-n \int_0^t dt' \gamma(\bar{g}(t'))\right\} \qquad (4.18)$$

but we usually prefer to write the integral in (4.18) in the form

$$\int_g^{\bar{g}(t)} dg' \frac{\gamma(g')}{\beta(g')} \qquad (4.19)$$

The gluon propagator has a gauge parameter ξ ($= 1, 0$ in Landau, Feynman gauges) so that the Callan–Symanzik equation (4.12) in QCD becomes

$$\left[\mu\frac{\partial}{\partial\mu} + \beta(g)\frac{\partial}{\partial g} - n_A\gamma_A(g) - n_F\gamma_F(g) + \delta(g)\frac{\partial}{\partial\xi}\right]$$
$$\times \Gamma^{(n_A,n_F)}(p, g, \xi, \mu) = 0 \quad (4.20)$$

where $\delta(g) = \mu \partial \xi / \partial \mu |_{g_0, \Lambda}$ and $\Gamma^{(n_A, n_F)}$ is the Green's function coupling n_A gluons and n_F quarks. The four functions $\beta(g), \delta(g), \gamma_A(g), \gamma_F(g)$ can be computed from a calculation of four independent Green's functions. A clear description of this procedure is given by Pennington (1983), the results being as follows:

$$\delta(g) = 2 \, \xi \, \gamma_A(g)$$

$$\gamma_F(g) = \frac{g^2}{16\pi^2} \, \xi \, C_2(F) \tag{4.21}$$

$$\gamma_A(g) = -\frac{g^2}{16\pi^2} \left[\left(\frac{13}{6} - \frac{\xi}{2} \right) C_2(A) - \tfrac{4}{3} N_F T_2(F) \right]$$

Taking the quark–gluon vertex $\Gamma^{(1,2)}$ and applying (4.20) gives

$$\beta(g) - \gamma_A(g)g - 2\gamma_F(g)g = -\frac{g^3}{16\pi^2} \left[\left(\frac{3}{2} + \frac{\xi}{2} \right) C_2(A) + 2\xi C_2(F) \right] \tag{4.22}$$

which gives $\beta(g) = -\beta_0(g^3/16\pi^2)$ with

$$\begin{aligned} \beta_0 &= \tfrac{11}{3} C_2(A) - \tfrac{4}{3} N_f T_2(F) \\ &= 11 - \tfrac{2}{3} N_f \end{aligned} \tag{4.23}$$

which means $\beta_0 > 0$ for less than 16 flavours. Thus (4.16) becomes, to this order,

$$\frac{d\bar{g}(t)}{dt} = \beta(\bar{g}) = -\beta_0 \frac{\bar{g}^3}{16\pi^2} \tag{4.24}$$

The strong coupling $\alpha_s(Q^2) = \bar{g}^2(Q^2)/4\pi$ and (4.24) can be written

$$\ln \frac{Q^2}{\mu^2} = -\int_{\alpha_s(\mu^2)}^{\alpha_s(Q^2)} \frac{4\pi}{\beta_0} \frac{d\alpha_s}{\alpha_s^2} \tag{4.25}$$

which gives

$$\frac{1}{\alpha_s(Q^2)} - \frac{\beta_0}{4\pi} \ln Q^2 = \frac{1}{\alpha_s(\mu^2)} - \frac{\beta_0}{4\pi} \ln \mu^2 \tag{4.26}$$

(4.26) can be satisfied only if both sides are equal to a constant; define this constant to be $-(\beta_0/4\pi) \ln \Lambda^2$ so that

$$\alpha_s(Q^2) = \frac{4\pi}{\beta_0 \ln (Q^2/\Lambda^2)} \tag{4.27}$$

The crucial aspect of this is that $\beta_0 > 0$ which gives the asymptotic freedom of the theory. That is, $\alpha_s(Q^2)$ has the behaviour shown

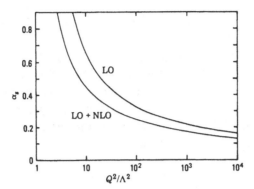

Fig. 4.2 The running coupling constant α_s of QCD computed for $N_f = 4$. The curves correspond to the leading order only and leading + next-to-leading order.

in fig. 4.2. This is to be contrasted with the behaviour seen in QED – an Abelian theory – where $\beta_0 < 0$. Notice that while the bare coupling g_0 is dimensionless, the renormalised coupling is now expressed in terms of a dimensionful quantity, Λ.

We see that the coupling constant α_s is characterised by this scale Λ, the value of which can be determined only from experiment. Intuitively, however, it would be surprising if Λ was very different from the scale which characterises the size of a typical hadron. For a proton, the radius is ~ 1 fm so we expect $\Lambda \sim 1$ fm $^{-1} \approx 200$ MeV.

From (4.24) we can write the solution for $\bar{g}(Q^2)$ in terms of $\bar{g}(\mu^2)$ as

$$\bar{g}(Q^2) = \bar{g}(\mu^2) \left[1 + \frac{\beta_0}{16\pi^2} \; \bar{g}^2(\mu^2) \ln \frac{Q^2}{\mu^2}\right]^{-\frac{1}{2}} \qquad (4.28)$$

Expanding this and comparing with (4.9) which is obtained by computing first the one-loop correction, then the two-loop terms and so on, we see that the RGE has the effect of summing the *leading logarithm* terms in (4.9). That is, the perturbative calculation of the coupling has been re-ordered so that the first term in the new series contains contributions from an *infinite* number of graphs in the old expansion. Perturbative expansions in QCD are therefore in terms of the running coupling $\alpha_s(Q^2)$ and so we expect perturbative calculations of physical quantities to be reliable if $\alpha_s(Q^2)$ is 'small' in contrast to a perturbative expansion in order of loops where we needed both $\alpha_s(\mu^2)$ and $\alpha_s(\mu^2) \ln(Q^2/\mu^2)$ to be small.

The next-to-leading order correction to $\beta(g)$ can be calculated (see chapter 5) and fig. 4.2 shows that the resulting change to the magnitude of $\alpha_s(Q^2)$ is felt at lower values of Q^2/Λ^2. These corrections are important however in giving a meaning to the magnitude of the QCD scale parameter Λ appearing in the definition of α_s, (4.27), as we shall see.

4.2   Regularisation

There are several procedures which can be adopted to renormalise the various Green's functions. That part of the procedure which involves deriving the precise finite intermediate expressions is called *regularisation*. The choice of counter terms is arbitrary and if we really could express a given physical quantity as an infinite series in powers of α_s then the value of this quantity would be independent of the choice of regularisation procedure. In real life we have only a truncated expansion and the magnitude of this finite series will depend on our specific choice.

Implicit in our discussion of renormalisation in the previous section was the concept of expressing the renormalised coupling at some large momentum Q^2 in terms of an 'assumed-to-be-known' coupling defined at some specific momentum scale μ^2. Essentially this regularisation procedure is the so-called momentum-subtraction (MOM) scheme. By defining $\bar{g}$ in terms of a vertex function where the external momenta are space-like with $p^2 = -m^2$ it turns out that a large class of higher order corrections can be absorbed into $\bar{g}$. This increases the possibility that the perturbative expansion will converge reasonably rapidly. The snags are that this procedure is actually gauge dependent and also that the results depend on the choice of the vertex, i.e. $q\bar{q}g$ or ggg.

An alternative way is the minimal-subtraction (MS) scheme which follows from the idea of dimensional regularisation ('t Hooft and Veltman, 1972). If we calculate a one-loop contribution to some Green's function we always obtain a factor

$$\int \frac{\mathrm{d}^4 k}{(2\pi)^4} \frac{\mathrm{i}}{k^2 - m^2 + i\epsilon} \qquad (4.29)$$

which, of course, diverges. Evaluating this integral in a space in which the number of dimensions is $n \neq 4$ gives a finite answer however. Writing $n = 4 - \epsilon$ then

$$\int \frac{\mathrm{d}^n k}{k^2 - m^2 + i\epsilon} \longrightarrow i\pi^{\frac{n}{2}} \, \Gamma(1 - \frac{n}{2}) \, (-m^2)^{\frac{n}{2}-1} \qquad (4.30)$$

with

$$\Gamma(1 - \frac{n}{2}) = -\frac{2}{\epsilon} - \tfrac{3}{2} + \gamma_E \qquad (4.31)$$

where γ_E is the Euler constant. The divergence of the loop now shows up explicitly as a pole in ϵ. When calculating Green's functions, we expand in ϵ and always get the singular piece $\sim \epsilon^{-1}$. The introduction of counter terms, designed to exactly cancel just this piece, defines the minimal subtraction (MS) scheme. If we want to also remove the terms involving γ_E then this defines a modified MS scheme or $\overline{MS}$ scheme. These subtraction schemes are popular since, unlike MOM schemes, the procedure is gauge invariant and easily generalised to multiloop graphs. Its drawback is that while g is dimensionless for n=4 it acquires dimensions for $n \neq 4$. So a scale μ enters into the procedure and factors of μ^ϵ keep appearing.

We shall see how the choice of scheme affects the definition of the basic scale parameter Λ, appearing in the expression for α_s, when we come to discuss the next-to-leading order corrections of the theory.

4.3 Operator product expansion

The key to applying QCD to DIS is the factorisation of the forward virtual Compton scattering amplitude into two parts: one which represents the long-distance physics for which perturbative techniques are useless and another which describes the light-cone physics. This latter factor has, in general, singularities which determine the large Q^2 behaviour of the amplitude and can be handled by the techniques of perturbative QCD.

Consider the forward scattering amplitude $T_{\mu\nu}(q^2, \nu)$ for a virtual photon scattering off a proton. To simplify the argument, let the photon be scalar so we can drop the μ, ν. This amplitude, shown in fig. 4.3 involves the time-ordered product of the (scalar)

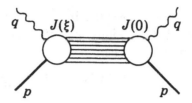

Fig. 4.3   Forward virtual photon–proton scattering amplitude

currents $J(\xi)$, $J(0)$:-

$$T(q^2,\nu) = \mathrm{i}\sum_\sigma \int \mathrm{d}^4\xi\, e^{iq\cdot\xi} <p,\sigma|T(J(\xi)\,J(0))|p,\sigma> \quad (4.32)$$

The amplitude is a real, analytic function of ν and q^2, i.e.

$$T(q^2,\nu^*) = T^*(q^2,\nu) \quad (4.33)$$

and is crossing symmetric

$$T(q^2,\nu) = T(q^2,-\nu) \quad (4.34)$$

There is a nucleon pole at $\nu = -q^2/2M$ and a cut from $\nu = (2Mm_\pi - q^2)/2M$ to $+\infty$, also a cut corresponding to the crossed process $p \to \gamma X$. The discontinuity across the right-hand cut is

$$\mathrm{Im}\, T(q^2,\nu) = 2\pi W(q^2,\nu) \quad (4.35)$$

where $W(q^2,\nu)$ is the familiar hadronic tensor (without the μ,ν !) and T satisfies the dispersion relation

$$T(q^2,\nu) = 4\int_{-q^2/2M}^{\infty} \frac{\mathrm{d}\nu'\nu'}{\nu'^2 - \nu^2}\, W(q^2,\nu) \quad (4.36)$$

We can write the r.h.s. in terms of $x = -q^2/2M\nu$,

$$T(q^2,\nu) = 4\sum_{n=2}^{\infty} \left(\frac{1}{x}\right)^n \int_{-1}^{1} \mathrm{d}x'\, x'^{n-1}\, W(q^2,x') \quad (4.37)$$

Our starting point is to write a Wilson (1969) operator product expansion of the time ordered products in (4.32) which have small light-cone separation (Brandt and Preparata 1971, Gross and Treiman 1971),

$$\mathrm{i}T(J(\xi)J(0))$$

$$\xrightarrow{\xi^2\to 0} \sum_{\tau=2}^{\infty}\sum_{n=0}^{\infty} C_{\tau,n}(\xi^2,\mu^2)\, \xi^{\mu_1}\dots\xi^{\mu_n}\, \mathcal{O}^{\tau}_{\mu_1\dots\mu_n}(\mu^2) \quad (4.38)$$

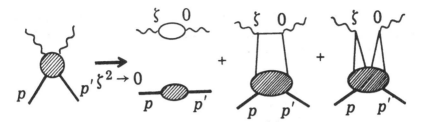

Fig. 4.4 Graphs generating light-cone singularities.

The operators $\mathcal{O}^\tau$ are chosen to be symmetric, traceless in all Lorentz indices. The Wilson coefficients $C_{\tau,n}$ are c-numbers, n being the spin of the operator $\mathcal{O}^\tau$, τ labels the *twist* (see below) and μ^2 is the renormalisation point.

Of course, different field theories have different operators and therefore different degrees of singularity for the $C_{\tau,n}$. For the simple example of a product of scalar currents, where interactions are neglected, we have

$$T(J(\xi)J(0))$$
$$= -2[\Delta_F(\xi,m^2)]^2 + 4i\Delta_F(\xi,m^2) : \phi(\xi)\phi(0) : + : \phi^2(\xi)\phi^2(0) :$$
$$\xrightarrow{\xi^2 \to 0} \frac{1}{8\pi^4(\xi^2 - i\epsilon)^2} - \frac{:\phi(\xi)\phi(0):}{\pi(\xi^2 - i\epsilon)} + : \phi^2(\xi)\phi^2(0) : \quad (4.39)$$

where Δ_F are the free propagators and : : means light-cone singularities are removed. Sandwiching (4.39) between proton states is shown in fig. 4.4. Thus (4.39) says that the coefficients have singularities of the form

$$C_\tau(\xi^2) \sim \left(\frac{1}{\xi^2}\right)^{d_{C_\tau}/2} \quad (4.40)$$

If d_J is the dimension of the current $J(\xi)$ and $d_\mathcal{O}$ is the dimension of the operator $\mathcal{O}$ then d_{C_τ} is given by

$$d_{C_\tau} = n - d_\mathcal{O} + 2d_J$$
$$= -\tau + 2d_J \quad (4.41)$$

where τ is called the *twist* of the operator. From (4.40) we see that those operators with *lowest* twist will be the most singular – hence the reason for ordering (4.38) in terms of twist. A quark current $\bar{q}\gamma_\mu q$ has twist = 2, as does the operator $\bar{q}\gamma^{\mu_1} D^{\mu_2} \ldots D^{\mu_n} q$.

The example above is particularly simple, there being no interactions. The numerator of the second term in (4.39) can be expanded in a Taylor series:

$$: \phi(\xi)\phi(0) : \; = \; \sum_n \frac{1}{2^n} \, \xi^{\mu_1} \dots \xi^{\mu_n} \; \phi(0) \; \overleftrightarrow{\partial}_{\mu_1} \dots \overleftrightarrow{\partial}_{\mu_n} \, \phi(0) \quad (4.42)$$

or – conversely – the series on the r.h.s. of (4.4.2) can be summed to give a bilocal operator. It turns out that when we go from this to an *interacting* theory, the coefficient functions will differ slightly for each term in the Taylor series and so prevent summation to give a simple bilocal operator. The consequence of this is to get *exact* scaling for a *free* field theory and *logarithmic* scaling for an interacting theory. Also in a free field theory the singular pieces (being just c-numbers) can be removed by normal ordering but in an interacting theory they can be removed only by renormalising the operators. That is why μ^2 appears in both the coefficient function and the operator in (4.38) even though the product must not depend on the choice of the renormalisation point.

Substituting the OPE (4.38) into (4.32) gives

$$T(q^2, \nu) \; \longrightarrow$$

$$\sum_{\tau, n} \frac{\partial}{\partial q_{\mu_1}} \dots \frac{\partial}{\partial q_{\mu_n}} \left[\int d^4\xi \, e^{iq \cdot \xi} \, C'_{\tau, n}(\xi^2, \mu^2) \left(\frac{1}{\xi^2}\right)^{1 - \frac{\tau}{2}} \right]$$

$$\times \, p_{\mu_1} \dots p_{\mu_n} \, \overline{O}_n^\tau(\mu^2) \qquad (4.43)$$

where

$$C_{\tau, n}(\xi^2, \mu^2) \; = \; C'_{\tau, n}(\xi^2, \mu^2) \left(\frac{1}{\xi^2}\right)^{1 - \frac{\tau}{2}}$$

and

$$< p| \, O^\tau_{\mu_1 \dots \mu_n}(\mu^2) \, |p > \; = \; \overline{O}_n^\tau(\mu^2) \, p_{\mu_1} \dots p_{\mu_n}$$

So we get

$$T(q^2, \nu) \; \longrightarrow \; \sum_{\tau, n} \left(\frac{2p \cdot q}{-q^2}\right)^n \overline{C}_{\tau, n}(Q^2, \mu^2) \left(\frac{1}{Q^2}\right)^{\frac{\tau}{2} - 1} \overline{O}_n^\tau(\mu^2)$$

$$\qquad (4.44)$$

$$= \; \sum_{\tau, n} \overline{C}_{\tau, n}(Q^2, \mu^2) \, \overline{O}_n^\tau(\mu^2) \left(\frac{1}{x}\right)^n \left(\frac{1}{Q^2}\right)^{\frac{\tau}{2} - 1} \qquad (4.45)$$

Only even values of n appear in this expansion since the amplitude is crossing symmetric. Therefore from (4.37) we can make the correspondence

$$\int_{-1}^{1} dx \, x^{n-1} \, W(q^2, x) = \frac{1}{4} \sum_{\tau=2}^{\infty} \overline{C}_{\tau,n}(Q^2, \mu^2) \, \overline{O}_n^{\tau}(\mu^2) \left(\frac{1}{Q^2}\right)^{\frac{\tau}{2}-1}$$
(4.46)

Now we go back and repeat the derivation with the vector current J_μ^{em}. The amplitude $T_{\mu\nu}$, just as the hadronic tensor $W_{\mu\nu}$, can be expressed in terms of two Lorentz scalars, giving

$$\int_{-1}^{1} dx \, x^{n-1} F_1(x, Q^2) = \overline{C}_{\tau=2,n}(Q^2, \mu^2) \, \overline{O}_n^{\tau=2}(\mu^2) + O\left(\frac{1}{Q^4}\right)$$
(4.47)

with a similar expression for F_2 except that we have x^{n-2} on the l.h.s. In this way we have singled out the precise term which describes the Q^2 dependence of the moments of the structure functions.

The physical interpretation of (4.47) is clearer if we transform to the structure function itself. If we define $f(y, \mu^2)$, $\sigma(z, Q^2, \mu^2)$ so that

$$\int dy \, y^{n-1} f(y, \mu^2) = \overline{O}_n^{\tau=2}(\mu^2)$$
$$\int dz \, z^{n-1} \sigma(z, Q^2, \mu^2) = \overline{C}_{\tau,n}(Q^2, \mu^2)$$
(4.48)

then (4.47) becomes a convolution integral

$$F_1(x, Q^2) = \int_x^1 \frac{dy}{y} f(y, \mu^2) \, \sigma(\frac{x}{y}, Q^2, \mu^2)$$
(4.49)

Thus $f(y, \mu^2)$ can be interpreted as the probability of finding a quark with momentum fraction y in the proton and $\sigma(z, Q^2, \mu^2)$ is the cross-section for that quark scattering elastically off a virtual photon with 'virtuality' Q^2. The radiation of gluons, as shown in fig. 4.5, carrying away momentum from that quark explains why $x \leq y$. From the graph of fig. 4.5 we understand the meaning of the scale μ^2. It simply represents the point in the ladder of gluons where we choose to separate into two components $f(y, \mu^2)$ and $\sigma(z, Q^2, \mu^2)$. Obviously this choice is quite arbitrary and the physical structure function (i.e. the product of the two components) is therefore independent of the scale μ^2.

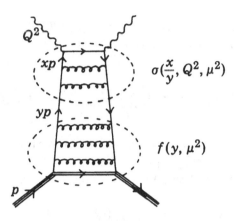

Fig. 4.5   Diagram for the leading logarithm, twist 2 contribution as given by (4.49).

4.4   An example: leading log behaviour of valence quarks

In a free field theory the coefficient functions $\overline{C}_{\tau,n}(Q^2,\mu^2) \rightarrow$ constant as $Q^2 \rightarrow \infty$ implying that the structure functions $F_i(x,Q^2)$ satisfy Bjorken scaling. In QCD, the interactions modify scaling by a logarithmic Q^2 dependence and the way to determine this precise Q^2 behaviour is to apply a scale transformation $q_\mu \rightarrow \lambda q_\mu$ and to carry out the renormalisation group analysis. This means computing the anomalous dimensions of the relevant twist = 2 operators.

Consider the flavour non-singlet operator, relevant to valence quarks. We express the anomalous dimensions of this operator as a series in powers of g^2,

$$\gamma^n(g) = \gamma_0^n \frac{g^2}{16\pi^2} + O(g^4) \qquad (4.50)$$

The only operator which carries flavour quantum numbers is given by

$$\mathcal{O}_n^{NS,\,\mu_1\cdots\mu_n} = \frac{i^{n-1}}{n!} \left[\bar{q}\, \frac{\lambda^a}{2}\, \gamma^{\mu_1}\, D^{\mu_2} \dots D^{\mu_n}\, q \; + \; \text{permutations} \right] \qquad (4.51)$$

To leading order, the relevant contributions come from the graphs of fig. 4.6 and these sum to give (Georgi and Politzer 1974, Gross

Fig. 4.6   Graphs contributing to $\gamma_0^{n,NS}$.

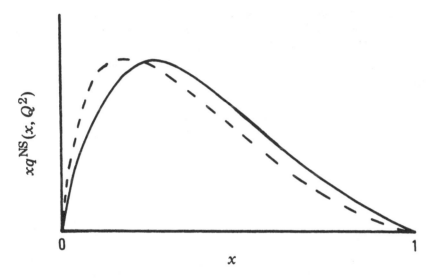

Fig. 4.7   Schematic behaviour of Q^2 dependence of non-singlet quark distribution.

and Wilczek 1974)

$$\gamma_0^{n,NS} = 2C_2(F) \left[1 - \frac{2}{n(n+1)} + 4 \sum_{j=2}^{n} \frac{1}{j} \right] \qquad (4.52)$$

Note that, because of the third term, $\gamma_0^n \sim \ln n$ for large n. We can now compute the solution to the RGE equation for the operator of (4.51).

$$\begin{aligned}
\overline{C}_n^{NS}\left(\frac{Q^2}{\mu^2}, g^2\right) &= \overline{C}_n^{NS}(1, \bar{g}^2(t)) \exp\left[-\int_0^t dt' \, \gamma^{n,NS}(\bar{g}) \right] \\
&= \overline{C}_n^{NS}(1, \bar{g}^2(t)) \exp\left[\frac{\gamma_0^{n,NS}}{\beta_0} \int_{\bar{g}(\mu^2)}^{\bar{g}(t)} \frac{d\bar{g}}{\bar{g}} \right] \qquad (4.53) \\
&= \overline{C}_n^{NS}(1, \bar{g}^2(t)) \, (\bar{g}^2)^{\gamma_0^{n,NS}/2\beta_0}
\end{aligned}$$

As a result, the moments of the non-singlet structure functions have a Q^2 dependence given by

$$M_n^{NS}(Q^2) = \int_0^1 dx\, x^{n-2}\, F_2^{NS}(x, Q^2) = A_n\, [\alpha_s(Q^2)]^{d_n^{NS}} \quad (4.54)$$

where $d_n^{NS} = \gamma_0^{n,NS}/2\beta_0$. From (4.52) we see that $d_{n=1}^{NS} = 0$, $d_n^{NS} > 0$ for $n > 1$ and $d_n^{NS} \to \ln n$ as $n \to \infty$. The pattern of scaling violations is established by (4.54). For $n > 1$ each moment $M_n^{NS}(Q^2)$ falls with Q^2 so that $F_2^{NS}(x, Q^2) = xq^{NS}(x, Q^2)$ falls with Q^2 at large x but rises at small x to maintain $\int dx q^{NS}(x, Q^2) = $ constant. The schematic behaviour is shown in fig. 4.7.

5

Applying QCD to deep inelastic scattering

We shall follow two approaches in deriving the Q^2 behaviour of structure functions in QCD. One, the more formal approach, continues the discussion of the last chapter involving the moments; the other is more intuitive where the Q^2 dependence is described by the Altarelli–Parisi equations. The two approaches are quite equivalent even to next-to-leading order which we discuss in some detail.

5.1 Moments of structure functions in leading order QCD

5.1.1 Non-singlet moments

In the last chapter we found the Q^2 dependence of the moments of the non-singlet structure functions was given by

$$
\begin{aligned}
M_n^{NS}(Q^2) &= \int \mathrm{d}x \, x^{n-2} \, F^{NS}(x, Q^2) \\
&= M_n^{NS}(Q_0^2) \, \left[\frac{\alpha_s(Q^2)}{\alpha_s(Q_0^2)} \right]^{d_n^{NS}}
\end{aligned}
\tag{5.1}
$$

where the $d_n^{NS} = \gamma_0^{n,NS}/2\beta_0$ are listed in table 5.1.

Conservation of flavour is guaranteed by $d_{n=1}^{NS} = 0$, consistent with the Gross–Llewellyn Smith and Bjorken sum rules. From (5.1) we see that

$$
Q^2 \frac{\mathrm{d}}{\mathrm{d}Q^2} \ln M_n(Q^2) = -\frac{\alpha_s(Q^2)}{4\pi} \, [\tfrac{1}{2}\gamma_0^{n,NS} + O(\alpha_s)]
\tag{5.2}
$$

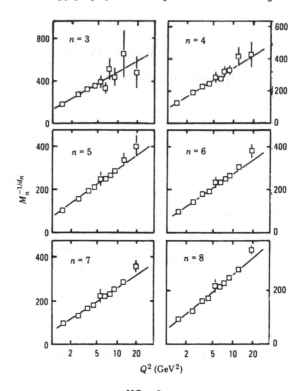

Fig. 5.1   Non-singlet moments $M_n^{NS}(Q^2)$ computed from muon and neutrino data, to the power $-1/d_n^{NS}$ against $\ln Q^2$. Taken from Pennington (1983).

Also, if we take two different values of n in (5.1) and eliminate α_s we get

$$\ln M_{n'}(Q^2) \;=\; \left(\frac{d_{n'}^{NS}}{d_n^{NS}}\right)\ln M_n(Q^2) \;+\; \text{const} \qquad (5.3)$$

So, if we plot, for various values of Q^2, one moment against another on a log-log plot, QCD predicts a straight line the slope of which is the ratio of the appropriate anomalous dimensions. Such plots (known as Perkins' plots) turn out not to be particularly sensitive measures of the predicted Q^2 behaviour – a more direct exercise is simply to plot the moment raised to the exponent $-1/d_n^{NS}$ against $\ln Q^2$ which, according to (5.1), will be linear.

From such plots, as in fig. 5.1, there is good evidence to support the expectation of (5.1) and in fact the intercept on the x-axis gives the value of $\ln \Lambda^2$ where Λ is the QCD scale parameter.

Such behaviour will clearly have corrections at low Q^2 from several sources. One source is higher order corrections (in α_s) and another is contributions from higher twist operators. In section 5.3 we shall discuss $O(\alpha_s)$ corrections which give additional terms in $\ln Q^2$ and the higher twist terms give extra powers of $1/Q^2$. Most of these are not calculable but some corrections of $1/Q^2$ type can be attributed to the non-zero mass of the target. These are discussed next.

5.1.2 Target mass corrections

Previously we saw that in the massless situation, only operators with spin $= n$ contribute to the nth moment. From (4.51) let us take an example of a spin 2 operator and form the matrix element for a proton target,

$$< p| \, \mathcal{O}^{NS}_{\mu\nu,n=2} \, |p> = \frac{i}{2!} < p| \, \bar{q}\frac{\lambda^a}{2}\gamma^\mu D^\nu q \; - \; \tfrac{1}{4}g_{\mu\nu}\bar{q}q \, |p> \quad (5.4)$$

The second term is required to make the operator traceless but clearly it gives rise to a term $-\tfrac{1}{4}M^2 g_{\mu\nu}$. As a result we get, in general, additional terms $\sim (M^2/Q^2)^n$ and many spins contribute to the nth moment. Nachtmann (1973) devised a way to project out the pure spin $= n$ contribution to the nth moment. For the F_2 structure function the result is

$$M_n(Q^2) = \int_0^1 \mathrm{d}x \, \frac{\xi^{n+1}}{x^3} \left[\frac{3 + 3(n+1)r + n(n+2)r^2}{(n+2)(n+3)} \right] F_2(x, Q^2) \tag{5.5}$$

where

$$r = \sqrt{\left(1 + \frac{4M^2 x^2}{Q^2}\right)} \quad , \quad \xi = 2x/(1+r) \tag{5.6}$$

and for F_3

$$M_n(Q^2) = \int_0^1 \mathrm{d}x \, \frac{\xi^{n+1}}{x^3} \left[\frac{1 + (n+1)r}{(n+2)} \right] x F_3(x, Q^2) \tag{5.7}$$

Given experimental measurements of $F_2, x F_3$, (5.5), (5.7) define the appropriate experimental quantities to which the QCD predictions of (5.1) should apply. What is the phenomenological effect of including these low Q^2 corrections? When $Q^2 > 4M^2 x^2$, (5.7) becomes

$$M_n(Q^2) \longrightarrow \int_0^1 dx \, x^{n-2} \left[1 - \frac{n(n+1)}{(n+2)} \frac{M^2 x^2}{Q^2}\right] x F_3(x, Q^2) \quad (5.8)$$

and, because of the minus sign, the Nachtmann moments fall less rapidly with Q^2 than the uncorrected moments. Consequently a lower value of Λ will result from an analysis of the data. However, if we take such low Q^2 seriously there is no reason to neglect the elastic contribution at $x = 1$. This contribution is relatively suppressed in the Nachtmann moment. As a result, taking both these low Q^2 considerations into account the net effect is to end up with practically no change to the original Q^2 dependence.

5.1.3 Singlet moments

The Q^2 dependence of the flavour singlet moments is more complicated since mixing can occur between the gluon and singlet-quark operators. In the flavour non-singlet case, we had just one set of operators, (4.51). The singlet-quark and gluon operators are

$$\mathcal{O}_n^{q,\mu_1\cdots\mu_n} = \frac{i^{n-1}}{n!} \left[\bar{q}\gamma^{\mu_1} D^{\mu_2} \ldots D^{\mu_n} q + \text{permutations}\right] \quad (5.9)$$

and

$$\mathcal{O}_n^{G,\mu_1\cdots\mu_n} = \frac{i^{n-2}}{n!} \, \text{Tr}\left[G^{\mu_1 \nu} D^{\mu_2} \ldots D^{\mu_{n-1}} G_\nu^{\mu_n} + \text{permutations}\right]$$

$$(5.10)$$

The solution to the RGE for the quark singlet and gluon coefficient functions takes the form, for $i = q, G$

$$\overline{C}_n^i \left(\frac{Q^2}{\mu^2}, g^2\right) = \sum_{j=q,G} \overline{C}_n^j (1, \bar{g}^2(t)) \exp\left[-\frac{\gamma_0^{n,ji}}{2\beta_0} \ln t\right] \quad (5.11)$$

where

$$\gamma^{n,ji}(g) = \gamma_0^{n,ji} \frac{g^2}{16\pi^2} + O(g^4) \quad (5.12)$$

The anomalous dimensions $\gamma_0^{n,ji}$ are calculated from one-loop graphs. The $\gamma_0^{n,ji}$ being obtained from the graphs of fig. 4.5 and the remaining ones from those of fig. 5.2.

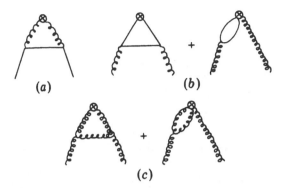

Fig. 5.2   Graphs contributing to (a) $\gamma_0^{n,Gq}$, (b) $\gamma_0^{n,qG}$, (c) $\gamma_0^{n,GG}$.

$\gamma_0^{n,qq}$ is equal to the non-singlet anomalous dimension $\gamma_0^{n,NS}$ to this order and the other diagonal element in the matrix is

$$\gamma_0^{n,GG} = 2C_2(A)\left[\frac{1}{3} - \frac{4}{n(n-1)} - \frac{4}{(n+1)(n+2)} + 4\sum_{j=2}^{n}\frac{1}{j}\right]$$

$$+ \tfrac{8}{3}N_f T_2(F) \tag{5.13}$$

Defining $\gamma_0^{n,\pm}$ as the eigenvalues of the anomalous dimension matrix, so that

$$\gamma_0^{\pm} = \tfrac{1}{2}\left\{(\gamma_0^{n,qq} + \gamma_0^{n,GG}) \pm [(\gamma_0^{n,qq} - \gamma_0^{n,GG})^2 + 4\,\gamma_0^{n,gG}\gamma_0^{n,Gq}]^{\frac{1}{2}}\right\} \tag{5.14}$$

we arrive at the final expression for the evolution of the quark singlet and gluon moments,

$$\begin{pmatrix} M_n^S(Q^2) \\ M_n^G(Q^) \end{pmatrix} =$$
$$\begin{pmatrix} (1-a_n)r_n^+ + a_n r_n^- & b_n(r_n^+ - r_n^-) \\ a_n(1-a_n)(r_n^+ - r_n^-)/b_n & a_n r_n^+ + (1-a_n)r_n^- \end{pmatrix} \begin{pmatrix} M_n^S(Q_0^2) \\ M_n^G(Q_0^2) \end{pmatrix} \tag{5.15}$$

where

$$a_n = \frac{\gamma_0^{n,qq} - \gamma_0^{n,+}}{\gamma_0^{n,-} - \gamma_0^{n,+}}, \qquad b_n = \frac{-\gamma_0^{n,qG}}{\gamma_0^{n,-} - \gamma_0^{n,+}} \tag{5.16}$$

and

$$r_n^\pm = \left[\frac{\alpha_s(Q^2)}{\alpha_s(Q_0^2)}\right]^{d_n^\pm}, \quad d_n^\pm = \gamma_0^{n,\pm}/2\beta_0 \qquad (5.17)$$

$M_n^S(Q^2)$ is the nth moment of the quark singlet structure function $\Sigma(x, Q^2)$

$$M_n^S(Q^2) = \int_0^1 dx \, x^{n-2} \, \Sigma(x, Q^2) \qquad (5.18)$$

While $M_n^S(Q^2)$ is clearly a physical quantity, $M_n^G(Q^2)$ is not. The ambiguity in its definition can be seen by transforming $M_n^G \to \lambda M_n^G, \gamma_0^{n,qG} \to \frac{1}{\lambda} \gamma_0^{n,qG}$ and $\gamma_0^{n,Gq} \to \lambda \gamma_0^{n,Gq}$ and inspecting that (5.15) is *invariant* under this transformation. That is, the off-diagonal elements of the anomalous dimension matrix depend on the normalisation of the gluon operator. If we want to normalise the $M_n^G(Q^2)$ such that the fractions of momentum carried by quarks and gluons add up to a constant independent of Q^2, then the appropriate convention is that of Gross and Wilczek (1974) where

$$\begin{aligned} \gamma_0^{n,qG} &= -8N_f \, T_2(F) \, \frac{(n^2+n+2)}{n(n+1)(n+2)} \\ \gamma_0^{n,Gq} &= -4C_2(F) \, \frac{(n^2+n+2)}{n(n^2-1)} \end{aligned} \qquad (5.19)$$

In this convention, we see that for $n = 2$

$$\begin{aligned} \gamma_0^{2,qq} + \gamma_0^{2,Gq} &= 0 \\ \gamma_0^{2,GG} + \gamma_0^{2,qG} &= 0 \end{aligned} \qquad (5.20)$$

So that (5.15) gives

$$M_2^S(Q^2) + M_2^G(Q^2) = M_2^S(Q_0^2) + M_2^G(Q_0^2) \qquad (5.21)$$

That is, energy-momentum is conserved and we take each side of (5.21) to be unity. The evolution of the $n=2$ singlet quark distribution takes on a simple form

$$\left[M_2^S(Q^2) - \frac{3N_f}{16 + 3N_f}\right] = \left[\frac{\alpha_s(Q^2)}{\alpha_s(Q_0^2)}\right]^{d_2^+} \left[M_2^S(Q_0^2) - \frac{3N_f}{16 + 3N_f}\right]$$

$$(5.22)$$

Thus the asymptotic values of the fractions of momentum carried

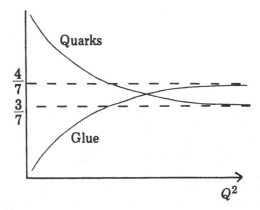

Fig. 5.3   Fractions of energy-momentum carried by quarks and gluons as a function of Q^2 for four flavours.

by quarks and glue are respectively $3N_f/(16+3N_f)$ and $16/(16+3N_f)$ which gives $\frac{3}{7}$ and $\frac{4}{7}$ for $N_f = 4$. The qualitative description of these fractions is shown in fig. 5.3.

Experiments indicate that for Q^2 in the range 10–40 GeV^2, each fraction is very close to $\frac{1}{2}$.

The actual values of the one-loop quantities we have used in this section are listed in table 5.1 (for the case $N_f = 4$). From the table we can see that as n increases, $a_n \to 1$, $b_n \to 0$, $a_n(1-a_n)/b_n \to 0$ and so for $n \gtrsim 6$ we can write

$$M_n^S(Q^2) \simeq r_n^- M_n^S(Q_0^2), \quad M_n^G(Q^2) \simeq r_n^+ M_n^G(Q_0^2) \qquad (5.23)$$

so that equations (5.15) effectively decouple at large n – i.e. large x behaviour of the quarks is *independent* of the gluon distribution.

Finally we notice that leading order QCD leaves the sum rules of the QPM (section 3.4) unchanged. Sum rules (a)–(d) are independent of Q^2 because $d_{n=1}^{NS} = 0$ and sum-rule (e) is constant because $d_{n=2}^- = 0$. Also to leading order F_L is zero and so the QPM result $R \equiv \sigma_L/\sigma_T = 0$ is maintained.

5.2   Altarelli–Parisi equations in leading order QCD

There is another way to derive the Q^2 dependence of the structure functions, based on an intuitive view (Kogut and Susskind 1974)

Table 5.1. Constants of asymptotic freedom for $N_f = 4$ (Gross-Wilczek convention)

n	d_n^{qq}	d_n^{qG}	d_n^{Gq}	d_n^{GG}	d_n^+	d_n^-	a_n	b_n
2	0.4267	−0.3200	−0.4267	0.3200	0.7467	0.0000	0.4286	−0.4286
3	0.6667	−0.2240	−0.1867	1.3280	1.3861	0.6085	0.9253	−0.2881
4	0.8373	−0.1760	−0.1173	1.8320	1.8523	0.8170	0.9803	−0.1700
5	0.9707	−0.1463	−0.0853	2.1817	2.1919	0.9604	0.9917	−0.1188
6	1.0804	−0.1257	−0.0670	2.4543	2.4604	1.0743	0.9956	−0.0907
7	1.1737	−0.1105	−0.0552	2.6794	2.6835	1.1697	0.9973	−0.0730
8	1.2550	−0.0987	−0.0470	2.8720	2.8749	1.2521	0.9982	−0.0608
9	1.3270	−0.0892	−0.0409	3.0406	3.0427	1.3249	0.9988	−0.0519
10	1.3916	−0.0815	−0.0362	3.1908	3.1924	1.3900	0.9991	−0.0452
11	1.4503	−0.0750	−0.0325	3.3263	3.3276	1.4490	0.9993	−0.0399
12	1.5040	−0.0695	−0.0295	3.4498	3.4508	1.5030	0.9995	−0.0356
13	1.5535	−0.0647	−0.0270	3.5633	3.5642	1.5527	0.9996	−0.0322
14	1.5995	−0.0606	−0.0248	3.6683	3.6691	1.5987	0.9996	−0.0293
15	1.6423	−0.0569	−0.0230	3.7661	3.7667	1.6417	0.9997	−0.0268
16	1.6825	−0.0537	−0.0215	3.8575	3.8581	1.6820	0.9998	−0.0247
17	1.7203	−0.0509	−0.0201	3.9434	3.9439	1.7198	0.9998	−0.0229
18	1.7559	−0.0483	−0.0189	4.0245	4.0249	1.7555	0.9998	−0.0213
19	1.7897	−0.0460	−0.0179	4.1011	4.1015	1.7894	0.9998	−0.0199
20	1.8218	−0.0438	−0.0169	4.1738	4.1741	1.8215	0.9999	−0.0187
21	1.8523	−0.0419	−0.0161	4.2430	4.2433	1.8520	0.9999	−0.0176
22	1.8815	−0.0402	−0.0153	4.3090	4.3093	1.8812	0.9999	−0.0165
23	1.9094	−0.0385	−0.0146	4.3721	4.3723	1.9091	0.9999	−0.0157
24	1.9361	−0.0370	−0.0140	4.4325	4.4327	1.9359	0.9999	−0.0148
25	1.9617	−0.0357	−0.0134	4.4905	4.4907	1.9615	0.9999	−0.0141

of the resolution of quarks in a nucleon as the wavelength of the probing virtual photon changes. This approach avoids the formal machinery of the RGE and OPE but is completely equivalent. Increasing the resolving power of the photon – i.e. increasing Q^2, means there is a greater likelihood for a single parton to appear as two, or more, partons.

Consider first the QPM as described by fig. 5.4(a). The quark density $q(x)$ is defined by

$$d\sigma(\gamma^* P \longrightarrow k'X) = \int dy\, q(y)\, d\sigma(\gamma^* k \longrightarrow k') \qquad (5.24)$$

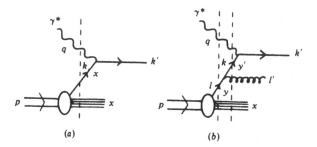

Fig. 5.4   Increasing Q^2 to $Q^2 + dQ^2$ increases probability of intermediate quark in (a) to appear as quark + gluon in (b).

The cross-sections in (5.24) are expressed in terms of scattering matrix elements

$$d\sigma(\gamma^*P \longrightarrow k'X) = \frac{1}{8E_P E_\gamma}|M(\gamma^*P \longrightarrow k'X)|^2$$

$$\times (2\pi)^4 \delta^4(q + P - k' - P_X)\frac{d^3 P_X}{(2\pi)^3 2E_X}\cdot\frac{d^3 k'}{(2\pi)^3 2E'} \quad (5.25)$$

$$d\sigma(\gamma^*k \longrightarrow k') = \frac{1}{8E_k E_\gamma}|M(\gamma^*k \longrightarrow k')|^2$$

$$\times (2\pi)^4 \delta^4(q + k - k')\frac{d^3 k'}{(2\pi)^3 2E'}$$

The contribution to $\gamma^*P \to k'X$ from the intermediate state shown as the dashed line in fig. 5.4(a) is given by

$$M(\gamma^*P \to k'X)$$
$$= M(\gamma^*k \to k')\cdot M(P \to kX)(E_X + E_k - E_P)^{-1}(2E_k)^{-1} \quad (5.26)$$

Thus the quark density is

$$q(y)dy = \frac{E_k}{E_P}|M(P \longrightarrow kX)|^2(E_X + E_k - E_P)^{-2}(2E_k)^{-2}\frac{d^3 P_X}{(2\pi)^3 2E_X} \quad (5.27)$$

Because $d\sigma(\gamma^*k \to k') = C_\gamma\delta(y - x)$ where $x = Q^2/2P\cdot q$ we can write

$$q(x) = C_\gamma^{-1}d\sigma(\gamma^*P \longrightarrow k'X) \quad (5.28)$$

where C_γ depends only on x, Q^2.

Now we turn to the emission of a gluon, fig. 5.4(b). Using the same approach we get

$$dq(x) = C_\gamma^{-1} d\sigma(\gamma^* P \longrightarrow k'l'X)$$

$$= \frac{E_l}{E_P}|M(P \longrightarrow lX)|^2(E_X + E_l - E_P)^{-2}(2E_l)^{-2}\frac{d^3P_X}{(2\pi)^3 2E_X}$$

$$\times \frac{E_k}{E_l}|M(l \longrightarrow kl')|^2(E_k + E_{l'} - E_l)^{-2}(2E_k)^{-2}\frac{d^3k_{l'}}{(2\pi)^3 2E_{l'}}$$

$$\times C_\gamma^{-1} d\sigma(\gamma^* k \longrightarrow k')$$

$$(5.29)$$

The first factor is $\int dy\, q(y)$, the third factor is just $\delta(y' - x)$, so

$$dq(x) = \int dy\, q(y)\left[\frac{E_k}{E_l}|M(l \longrightarrow kl')|^2(2E_k)^{-2}\right.$$

$$\left.\times(E_k + E_{l'} - E_l)^{-2}\frac{d^3k_{l'}}{(2\pi)^3 2E_{l'}}\right]\delta(y' - x)$$

$$(5.30)$$

Writing the components of the momenta as

$$l = (yP,\ \mathbf{0},\ yP)$$
$$k = (y'P + l_T^2/2yP,\ \mathbf{l_T},\ y'P)$$
$$l' = ((y - y')P + l_T^2/2(y - y')P,\ -\mathbf{l_T},\ (y - y')P)$$

and putting $z = y'/y$ we get

$$(E_k + E_{l'} - E_l)^2 = [l_T^2/z(1 - z)2yP]^2$$
$$(8E_l E_k E_{l'}) = [8(yP)^3 z(1 - z)]$$
$$d^3k_{l'} = \pi\, yP\, dz\, dl_T^2$$
$$|M(l \longrightarrow kl')|^2 = \frac{1}{2}g^2 C_2(F)\ \text{Tr}[\gamma \cdot l\gamma_\mu\gamma \cdot k\gamma_\nu]\sum_{pol}\epsilon^{*\mu}(l')\epsilon^\nu(l')$$

$$= 2g^2 C_2(F)(k_\mu l_\nu + l_\mu k_\nu - g_{\mu\nu}l \cdot k)\left[\delta^{ij} - \frac{l'^i l'^j}{l'^2}\right]$$

$$= 8\pi\alpha_s C_2(F)\frac{1 + z^2}{z(1 - z)^2}l_T^2$$

Substituting all these terms into (5.30) then gives

$$dq(x) = \int dy\, q(y)\left[\frac{\alpha_s}{2\pi}P_{qq}(z)\, d\ln l_T^2\, dz\right]\delta(zy - x) \qquad (5.31)$$

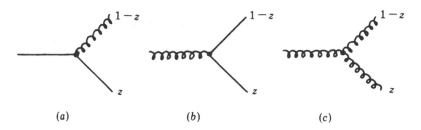

Fig. 5.5 Vertices relevant to parton splitting functions. (a) Quark-gluon vertex which determines P_{qq} and P_{Gq}. (b) Annihilation vertex of $G \to q\bar{q}$ which determines P_{Gq}. (c) Three gluon vertex which determines P_{GG}.

where

$$P_{qq}(z) = C_2(F) \frac{1 + z^2}{1 - z} \qquad (5.32)$$

is the *splitting function* which determines the probability for a quark radiating a gluon such that the quark's momentum is reduced by a fraction z, fig. 5.5(a).

The expression for $dq(x)$ given by (5.31) is positive but this is not the whole contribution. The emission of the gluon in fig. 5.4(b) arises from the increase in Q^2 – it is this scale which sets the size of the proton transverse momentum l_T^2. When the scale changes from Q^2 to $Q^2 + dQ^2$ there are two sources of the change in quark density (Collins and Qiu 1988). A positive contribution or 'gain' arises when the 'new' quark with momentum fraction x is obtained after the quark with momentum fraction y at the scale Q^2 emits a gluon ($y > x$). This is $dq(x)$ given by (5.31). There is also a negative contribution arising from the quarks that had momentum fraction x before radiating a gluon with fraction less than x. It is the same physical mechanism that is responsible and so has the same splitting function. Together these contributions give

$$\frac{dq^{NS}(x, Q^2)}{d \ln Q^2} = \frac{\alpha_s(Q^2)}{2\pi}$$

$$\times \int_0^1 dy \, dz \, q^{NS}(y, Q^2) \, P_{qq}(z) \, [\delta(zy - x) - \delta(y - x)] \quad (5.33)$$

This equation describes the evolution of the non-singlet quark density, and it is trivial to see that the number of valence quarks is automatically conserved. Equation (5.33) is the non-singlet version of the Altarelli-Parisi (1977) equations which were also derived by Dokshitzer, Dyakonov and Troyan (1978, 1980). The more familiar form of the A-P equation is obtained by absorbing the second δ function into the definition of $P_{qq}(z)$,

$$P_{qq}(z) \longrightarrow C_2(F) \left[\frac{(1+z^2)}{(1-z)_+} - \frac{3}{2}\delta(1-z) \right] \qquad (5.34)$$

where the regularised function $(1-z)_+^{-1}$ is defined by

$$\int_0^1 dz \frac{f(z)}{(1-z)_+} = \int_0^1 dz \frac{f(z)-f(1)}{(1-z)} = \int_0^1 dz \ln(1-z)\frac{df(z)}{dz}$$

The non-singlet evolution equation can then be written as

$$\frac{dq^{NS}(x,Q^2)}{d\ln Q^2} = \frac{\alpha_s(Q^2)}{2\pi} \int_x^1 \frac{dy}{y} q^{NS}(y,Q^2) P_{qq}(\frac{x}{y}) \qquad (5.35)$$

Defining moments, $M^{NS}(Q^2) = \int_0^1 dx\, x^{n-1} q^{NS}(x,q^2)$, (5.35) gives

$$Q^2 \frac{dM_n^{NS}(Q^2)}{dQ^2} = \frac{\alpha_s(Q^2)}{2\pi} \int_0^1 dz\, z^{n-1} P_{qq}(z) \qquad (5.36)$$

and comparing with (5.2) gives the connection between the *splitting function* $P_{qq}(z)$ and the anomalous dimension $\gamma_0^{n,NS}$,

$$\int_0^1 dz\, z^{n-1} P_{qq}(z) = -\frac{1}{4} \gamma_0^{n,NS} \qquad (5.37)$$

So far we have considered only one flavour. Generalising to all flavours and including the gluon contribution gives

$$\frac{dq^i(x,Q^2)}{d\ln Q^2} = \frac{\alpha_s(Q^2)}{2\pi}$$
$$\times \int_x^1 \frac{dy}{y} \left[\sum_j q^j(y,Q^2) P_{q^i q^j}(\frac{x}{y}) + G(y,Q^2) P_{q^i G}(\frac{x}{y}) \right]$$

$$\frac{dG(x,Q^2)}{d\ln Q^2} = \frac{\alpha_s(Q^2)}{2\pi}$$
$$\times \int_x^1 \frac{dy}{y} \left[\sum_j q^j(y,Q^2) P_{Gq^j}(\frac{x}{y}) + G(y,Q^2) P_{GG}(\frac{x}{y}) \right]$$
$$(5.38)$$

where i and j run over quarks *and* antiquarks of *all* flavours. Several simplifications are obvious however. At this order, the quark lines of fig. 5.5(a) cannot change flavour and so

$$P_{q^i q^i} = \delta_{ij} P_{qq}$$

Also the probability that a quark emitting a gluon is independent of flavour,

$$P_{Gq^j} = P_{Gq}$$

and a gluon creates a massless $q\bar{q}$ with a probability independent of flavour,

$$P_{q^i G} = P_{qG}$$

So we can write the Altarelli–Parisi (AP) equations for the quark singlet and gluon distributions in the form

$$\frac{\mathrm{d}}{\mathrm{d}\ln Q^2} \begin{pmatrix} q^S(x,Q^2) \\ G(x,Q^2) \end{pmatrix} =$$

$$\frac{\alpha_s(Q^2)}{2\pi} \int_x^1 \frac{\mathrm{d}z}{z} \begin{pmatrix} P_{qq}(z) & N_f P_{qG}(z) \\ P_{Gq}(z) & P_{GG}(z) \end{pmatrix} \begin{pmatrix} q^S(\frac{x}{z},Q^2) \\ G(\frac{x}{z},Q^2) \end{pmatrix}$$
$$(5.39)$$

and the generalisation of (5.37) is

$$\int_0^1 \mathrm{d}z\, z^{n-1} P_{ij}(z) = -\tfrac{1}{4}\, \gamma_0^{n,ij} \qquad (5.40)$$

That is, the splitting functions are the inverse Mellin transforms of the anomalous dimensions encountered in section 5.1, and we can use (5.40) to compute each of the P_{ij} using the expressions (4.50), (5.13) and (5.19).

Conservation of momentum for a parent quark and gluon respectively gives

$$\int_0^1 \mathrm{d}z\, z \left[P_{qq}(z) + P_{Gq}(z) \right] = 0$$
$$(5.41)$$
$$\int_0^1 \mathrm{d}z\, z \left[2N_f P_{qG}(z) + P_{GG}(z) \right] = 0$$

and also gives for $z < 1$

$$P_{qq}(z) = P_{Gq}(1-z), \quad P_{qG}(z) = P_{qG}(1-z), \quad P_{GG}(z) = P_{GG}(1-z)$$
$$(5.42)$$

The first equality gives, using (5.32),

$$P_{Gq}(z) = C_2(F)\, \frac{1 + (1-z)^2}{z} \qquad (5.43)$$

Calculating P_{qG}, P_{GG} for $z < 1$ gives

$$P_{qG}(z) = T_2(F)\, [z^2 + (1-z)^2] \qquad (5.44)$$

and

$$P_{GG}(z) = 2C_2(A)\left[\frac{z}{(1-z)_+} + \frac{1-z}{z} + z(1-z)\right] + \tfrac{1}{2}\beta_0\delta(1-z)$$

(5.45)

where β_0 is given by (4.23)

We can express the evolution of the non-singlet structure function $F^{NS}(x,Q^2) = xq^{NS}(x,Q^2)$. From the definition of $(1-z)_+$ we can write

$$\int_x^1 dz f(\tfrac{x}{z})(1-z)_+^{-1} = \int_0^1 dz f(\tfrac{x}{z})(1-z)_+^{-1}$$
$$- \int_0^x dz f(\tfrac{x}{z})(1-z)_+^{-1}$$
$$= \int_x^1 dz \left[f(\tfrac{x}{z}) - f(x)\right](1-z)^{-1}$$
$$- f(x)\int_0^x dz(1-z)^{-1}$$

and get

$$\frac{dF^{NS}(x,Q^2)}{d\ln Q^2} = \frac{\alpha_s(Q^2)}{2\pi} C_2(F)\left\{[\tfrac{3}{2} + 2\ln(1-x)]F^{NS}(x,Q^2)\right.$$
$$\left. + \int_x^1 \frac{dz}{(1-z)}[(1+z^2)\,F^{NS}(\tfrac{x}{z},Q^2) - 2F^{NS}(x,Q^2)]\right\}$$

(5.46)

The second term is negative for all x while the first term is positive for $x > 0.53$. For a typical non-singlet distribution the two terms cancel for $x \simeq 0.2$ and so scaling occurs. From figs. 5.6 and 5.7 we see that, in general, $dF_2/d\ln Q^2 = 0$ for x close to 0.2.

The A-P equations provide a practical way of determining the value of Λ_{QCD} from a set of experimental data. Suppose the set consists of values of $xF_3^{\nu N}$ over a range of x and Q^2. Then (5.46) can be used to analyse the data with a set of parameters which describe the structure function at some value of $Q^2 = Q_0^2$ together with Λ. For a general F_2, equations (5.39) are used and so a parametrisation of the gluon at Q_0^2 is also needed. The shape of the assumed gluon distribution is very hard to fix from such an analysis. The reason is that when we consider $dq^S/d\ln Q^2$ in (5.39), the second term (involving P_{qG}) is positive and for $x \gtrsim 0.25$ the first term (involving P_{qq}) is negative. So if we make the gluon

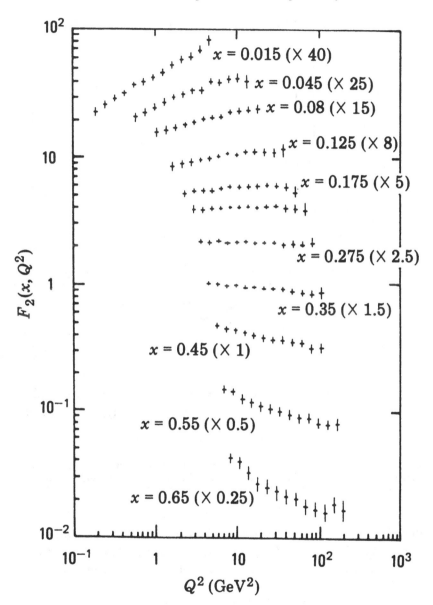

Fig. 5.6 The structure function $F_2^{\nu N}$ from the CDHSW collaboration (Berge *et al.* 1989).

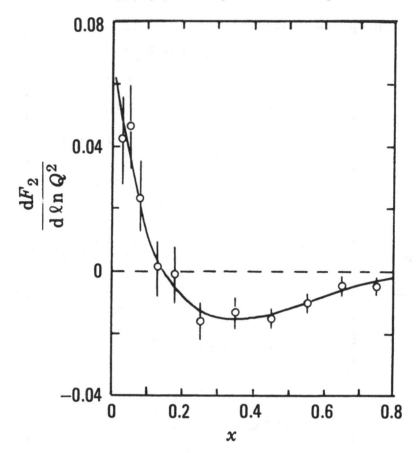

Fig. 5.7   Derivative of the structure function F_2 measured in muon-proton scattering by EMC (Aubert *et al.* 1986). The curve corresponds to a leading order fit with $\Lambda = 90$ MeV.

distribution 'harder' the l.h.s. of (5.39) can be maintained by increasing the values of α_s, i.e. increasing Λ. Thus it is difficult to pin down the starting gluon distribution and the value of Λ simultaneously.

5.2.1   Target mass corrections

Earlier we discussed the modification to moments to take account of the finite mass of the target. Here we discuss the question of the appropriately modified structure function which should be inserted into the A-P equations.

Instead of selecting only the spin $= n$ contribution to form the nth Nachtmann moment one can alternatively write the contribution to the usual moment and get (Georgi and Politzer 1976)

$$\int_0^1 \mathrm{d}x x^{n-2} F_2(x, Q^2) = \sum_{j=0}^\infty \left(\frac{M^2}{Q^2}\right)^j \frac{(n+j)!}{j!(n-2)!(n+2j)(n+2j-1)}$$

$$\times \sum_{i,\tau} \left(\frac{1}{Q^2}\right)^{\frac{\tau}{2}-1} \overline{C}^i_{\tau, n+2j}(Q^2) \overline{O}^i_{\tau, n+2j}$$

$$(5.47)$$

The twist $= 2$ contribution is defined as the moment of an auxilliary structure function $\mathcal{F}_2(x, Q^2)$, i.e.

$$\sum_i \overline{C}^i_{2,n}(Q^2) \overline{O}^i_{2,n} = \int_0^1 \mathrm{d}x\, x^n \mathcal{F}_2(x, Q^2) \qquad (5.48)$$

The normal structure function is usually expressed (De Rujula, Georgi and Politzer 1977) in terms of this function as

$$F_2(x, Q^2) = \frac{x^2}{r^3} \mathcal{F}_2(\xi, Q^2) + 6 \frac{M^2 x^2}{Q^2 r^4} \int_\xi^1 \mathrm{d}\xi' \mathcal{F}_2(\xi, Q^2)$$

$$+ 12 \frac{M^4 x^4}{Q^4 r^5} \int_\xi^1 \mathrm{d}\xi' \int_{\xi'}^1 \mathrm{d}\xi'' \mathcal{F}_2(\xi'', Q^2)$$

$$(5.49)$$

As in section 5.1, $\xi = 2x/(1+r)$ where $r^2 = 1 + 4M^2 x^2/Q^2$ so that $\xi = \xi_{th} < 1$ when $x = 1$. Thus one defect of (5.49) is that there is clear mismatch when $x = 1$ between the l.h.s. (which vanishes) and the r.h.s. (which does not). In fact (5.49) cannot be correct for we can write (5.47) in the form (Miramontes and Sanchez Guillen 1988)

$$\int_0^1 \mathrm{d}x x^{n-2} F_2(x, Q^2) = \int_0^{1/(1-M^2/Q^2)} \mathrm{d}x x^n \frac{\partial}{\partial x^2} \left[\frac{x^2 H(\xi, Q^2)}{\xi^2 r}\right]$$

$$(5.50)$$

where

$$H(x, Q^2) = \int_x^1 \mathrm{d}y\, (y - x)\, \mathcal{F}(y, Q^2) \qquad (5.51)$$

The *integrand* on the r.h.s. of (5.50) is the r.h.s. of (5.49) so the difference between the two upper limits of x on each side of (5.50) demonstrates (5.49) must be wrong. The conclusion is that the use of exact Nachtmann moments $M_n(Q^2)$ is formally incompatible with the notion of considering only twist $= 2$ contributions. One cannot approximate

$$M_n(Q^2) = \sum_{i,\tau} \left(\frac{1}{Q^2}\right)^{\frac{\tau}{2}-1} \overline{C}^i_{\tau,n}(Q^2) \overline{O}^i_{\tau,n} \qquad (5.52)$$

by (5.48), the pure twist $= 2$ contribution being at most logarithmically dependent on Q^2. Making such an approximation simply induces non-physical thresholds into the structure function (Gross, Treiman and Wilczek 1977). Even introducing prescriptions to partially reproduce these thresholds (Bitar, Johnson and Tung 1979) is probably not suitable since it must correspond to summing only *part* of the higher order terms. The problems can, in principle, only be resolved by including explicit higher twist terms (Miramontes and Sanchez Guillen 1988) and we shall return to this topic in chapter 6.

5.3 NLO corrections to moments of structure functions

It is easy to see that changing the value of Λ, e.g. $\Lambda \to \Lambda' = \Lambda/\kappa$, leaves (5.2) unchanged simply because κ can be absorbed into the $O(\alpha_s)$ correction. It is therefore important to compute the next-to-leading-order (NLO) corrections to define the scale in the theory.

Firstly, the two-loop corrections to $\beta(g)$ are needed. Writing the expansion as

$$\frac{d}{d \ln Q^2}\left(\frac{\alpha_s(Q^2)}{4\pi}\right) = -\beta_0 \left(\frac{\alpha_s(Q^2)}{4\pi}\right)^2 - \beta_1 \left(\frac{\alpha_s(Q^2)}{4\pi}\right)^3 \dots \quad (5.53)$$

with the one-loop term β_0 given by (4.23), the two-loop term has been computed as (Caswell 1974, Jones 1974)

$$\begin{aligned} \beta_1 &= \tfrac{34}{3}C_2(A)^2 - \tfrac{20}{3}C_2(A)T_2(F)N_f - 4C_2(F)T_2(F)N_f \\ &= 102 - \tfrac{38}{3}N_f \end{aligned} \qquad (5.54)$$

using $C_2(A) = 3$, $T_2(F) = \tfrac{1}{2}$, $C_2(F) = \tfrac{4}{3}$ in $SU(3)$. We can then compute the running coupling constant including this term. Putting $a = \beta_0\alpha_s(Q^2)/4\pi$, (5.53) becomes

$$\frac{da}{dt} = -a^2 - ba^3 \qquad (5.55)$$

with $b = \beta_1/\beta_0^2$, $t = \ln(Q^2/\mu^2)$. Instead of (4.26) we now get

$$\frac{1}{a} - b \ln\left(\frac{1 + ba}{a}\right) - \ln Q^2 = \frac{1}{a(\mu^2)} - b \ln\left(\frac{1 + b\,a(\mu^2)}{a(\mu^2)}\right) - \ln \mu^2$$

and, parallel to our leading order argument, we put each side equal to $-\ln \Lambda^2$ and so obtain

$$\ln \frac{Q^2}{\Lambda^2} = \frac{4\pi}{\beta_0 \alpha_s} - \frac{\beta_1}{\beta_0^2} \ln\left[\frac{4\pi}{\beta_0 \alpha_s} + \frac{\beta_1}{\beta_0^2}\right] \qquad (5.56)$$

For α_s small, we can invert this to get the approximation

$$\frac{\alpha_s(Q^2)}{4\pi} = \frac{1}{(\beta_0 \ln Q^2/\Lambda^2)} - \frac{\beta_1}{\beta_0} \frac{\ln \ln Q^2/\Lambda^2}{(\beta_0 \ln Q^2/\Lambda^2)^2} + \cdots \qquad (5.57)$$

Let us change the value of Λ once again, $\Lambda \rightarrow \Lambda' = \Lambda/\kappa$ and so the change in α_s is

$$\alpha_s \longrightarrow \alpha_s' = \alpha_s - \left[\frac{\beta_0}{4\pi} \ln \kappa\right] \alpha_s^2 + O(\alpha_s^3) \qquad (5.58)$$

Since α_s is defined within a given renormalisation prescription, going from one convention to another will therefore be equivalent to changing the magnitude of Λ.

5.3.1 *Non-singlet moments*

Calculating the NLO behaviour of the non-singlet moments is a matter of computing the quantities $\gamma_1^{n,NS}$ and B_n^{NS} which appear in the expansion of the anomalous dimension and the coefficient function,

$$\gamma^{n,NS} = \gamma_0^{n,NS} \left(\frac{\alpha_s}{4\pi}\right) + \gamma_1^{n,NS} \left(\frac{\alpha_s}{4\pi}\right)^2 + \cdots \qquad (5.59)$$

$$\overline{C}_n^{NS}(1, \bar{g}^2) = 1 + B_n^{NS} \left(\frac{\alpha_s}{4\pi}\right) + \cdots \qquad (5.60)$$

Calculating these two-loop corrections was an immense task; here we simply write down the results in each case. The values of the B_n^{NS} depend on the particular structure function. The result for the longitudinal structure function is simple (Zee, Wilczek and Treiman 1974).

$$B_{L,n}^{NS} = C_2(F)\frac{4}{n+1} \qquad (5.61)$$

Table 5.2. Values of the two-
loop quantities for non-singlet
moments.

n	$B_{2,n}^{NS}$	$\overline{B}_{2,n}^{NS}$	$\frac{\gamma_1^n}{2\beta_0}$	$-\frac{\gamma_0^n\beta_1}{2\beta_0^2}$
1	0.00	0.00	0.00	0.00
2	7.39	0.44	4.28	−2.63
3	14.08	3.22	6.05	−4.11
4	19.70	6.07	7.21	−5.16
5	24.53	8.73	8.09	−5.98
6	28.77	11.18	8.82	−6.66
7	32.55	13.44	9.44	−7.23
8	35.96	15.53	9.99	−7.73
9	39.08	17.48	10.46	−8.17
10	41.96	19.30	10.91	−8.57
11	44.63	21.01	11.31	−8.93
12	47.12	22.63	11.68	−9.27

and the relation between the NLO coefficients for F_2 and F_3 is
simple

$$B_{3,n}^{NS} = B_{2,n}^{NS} - C_2(F)\,\frac{(4n+2)}{n(n+1)} \qquad (5.62)$$

The actual values of $B_{2,n}^{NS}$, $B_{3,n}^{NS}$ depend on the choice of renormalisation scheme but $B_{L,n}^{NS}$ does not. In the $\overline{MS}$ scheme the result
is (Bardeen, Buras, Duke and Muta 1978)

$$B_{2,n}^{NS} = C_2(F)\left[3\sum_{j=1}^{n}\frac{1}{j} - 4\sum_{j=1}^{n}\frac{1}{j^2} - \frac{2}{n(n+1)}\sum_{j=1}^{n}\frac{1}{j}\right.$$

$$\left. + 4\sum_{s=1}^{n}\frac{1}{s}\sum_{j=1}^{s}\frac{1}{j} + \frac{3}{n} + \frac{4}{(n+1)} + \frac{2}{n^2} - 9\right] \qquad (5.63)$$

Notice that for large n, the fourth term in (5.63) dominates,
implying that $B_{2,n}^{NS} \sim \ln\ln n$. The calculation of the $\gamma_1^{n,NS}$
(Floratos, Ross and Sachrajda 1977) results in lengthy expressions
although more compact expressions have been found (Gonzalez-
Arroyo, Lopez and Yndurain 1979), but we list the numerical
values in table 5.2.

So using (5.59), (5.60) we can write the evolution of the non-singlet moments in the form

$$M_n^{NS}(Q^2) = M_n^{NS}(Q_0^2) \left[\frac{\alpha_s(Q^2)}{\alpha_s(Q_0^2)} \right]^{d_n}$$
$$\times \left[1 + C_n^{NS} \left(\frac{\alpha_s(Q^2) - \alpha_s(Q_0^2)}{4\pi} \right) \right] \quad (5.64)$$

where $d_n = \gamma_0^{n,NS}/2\beta_0$ as before and

$$C_n^{NS} = B_n^{NS} + \frac{\gamma_1^{n,NS}}{2\beta_0} - \frac{\beta_1 \gamma_0^{n,NS}}{2\beta_0^2} \quad (5.65)$$

The NLO term depends on the renormalisation scheme, e.g. the relation between them in the MS and $\overline{MS}$ schemes is

$$C_{n,\overline{MS}}^{NS} = C_{n,MS}^{NS} - \tfrac{1}{2}\gamma_0^n(\ln 4\pi - \gamma_E) \quad (5.66)$$

The logarithmic derivative of the moment is now given by

$$\frac{\mathrm{d}(\ln M_n^{NS}(Q^2))}{\mathrm{d}(\ln Q^2)}$$
$$= -\frac{\gamma_0^{n,NS}}{2} \left(\frac{\alpha_s}{4\pi} \right) - \left[\beta_0 C_n^{NS} + \frac{\beta_1 \gamma_0^{n,NS}}{2\beta_0} \right] \left(\frac{\alpha_s}{4\pi} \right)^2 + \dots \quad (5.67)$$

Of course the l.h.s. is a physical quantity and should be independent of the choice of renormalisation scheme. But since we are computing only a truncated perturbative expansion we can only insist on the scheme independence being valid to $O(\alpha_s^2)$. Thus if we change schemes so that

$$C_n^{NS} \longrightarrow C_n^{NS'} = C_n^{NS} + \gamma_0^n \ln \kappa \quad (5.68)$$

for some κ then the l.h.s. will remain invariant to $O(\alpha_s^2)$ provided that

$$\alpha_s \longrightarrow \alpha_s' = \alpha_s - (\frac{\beta_0}{2\pi} \ln \kappa) \alpha_s^2 \quad (5.69)$$

But from (5.58) we see that κ is the change in Λ which induces the transformation (5.69). Thus we can characterise a *change* of renormalisation scheme by the ratio Λ'/Λ. In case of going from the MS to $\overline{MS}$ scheme we see from (5.66), (5.69) that, in this case, $\kappa = \exp[-\tfrac{1}{2}(\ln 4\pi - \gamma_E)] = 1/2.66$ so that

$$\Lambda_{\overline{MS}} = 2.66 \, \Lambda_{MS} \quad (5.70)$$

It is important to realise that a change of renormalisation scheme affects only $O(\alpha_s^2)$ and since the β-function already starts at $O(\alpha_s^2)$ then the quantities β_0 and β_1 are independent of the choice of scheme. Higher terms are scheme dependent however.

The $\overline{MS}$ scheme appears to be preferable to the MS scheme insofar that the C_n^{NS} are smaller in the former and would suggest that the perturbative series is apparently converging more rapidly. Carrying this argument one step further we can arrange to go to a 'scheme' where the new coefficient C_n^{NS} is identically zero. To achieve this we take $\kappa = \exp(-C_{n,\overline{MS}}^{NS}/\gamma_0^{n,NS})$ so that $\Lambda' = \Lambda_n$ where

$$\Lambda_n = \Lambda_{\overline{MS}} \exp[C_{n,\overline{MS}}^{NS}/\gamma_0^{n,NS}] \tag{5.71}$$

and then the evolution of the non-singlet moments is just the leading order form,

$$M_n^{NS}(Q^2) = M_n^{NS}(Q^2) \left[\frac{\alpha_{s,n}(Q^2)}{\alpha_{s,n}(Q_0^2)}\right]^{d_n} \tag{5.72}$$

where $\alpha_{s,n}$ is the leading order coupling in terms of Λ_n. The attraction of the 'Λ_n scheme' (Bardeen *et al.* 1978) is that the NLO corrections appear in a rather graphic way. Computing the moments from data and fitting each separately to the leading order formula gives a set of values of Λ and if these agree with the monotonic increase given by (5.71) then there is evidence for NLO corrections. See the analysis by Duke, Owens and Roberts (1982), for example.

The question of the most suitable choice of renormalisation scheme is a vexed subject (Stevenson 1981a,1981b, Pennington 1982). Here we just mention that, in the context of DIS, the commonly employed schemes are MS, $\overline{MS}$ and MOM (Landau gauge), the connection between the latter pair being $\alpha_{\overline{MS}} = \alpha_{MOM}(1 - C_2(A)\frac{187}{144}(\alpha_{MOM}/\pi))$, so we get

$$\Lambda_{MOM} = 2.55 \, \Lambda_{\overline{MS}} \tag{5.73}$$

The problem of scheme dependence can be avoided by devising quantities which are scheme independent. Notice from the discussion above that the combination $K_{mn} = C_n^{NS}/\gamma_0^{n,NS} - C_m^{NS}/\gamma_0^{m,NS}$ is invariant under a change of scheme. Thus the Perkins' plots discussed in section 5.1 form a 'good' prediction (Pennington and Ross 1979) of QCD since the slope of one moment plotted against another on a log-log plot is

$$\frac{\mathrm{d}[\ln M_n(Q^2)]}{\mathrm{d}[\ln M_m(Q^2)]} = \frac{\gamma_0^{n,NS}}{\gamma_0^{m,NS}}\left\{1 + \frac{K_{mn}}{\ln Q^2/\Lambda^2}\right\} \qquad (5.74)$$

5.3.2   QCD corrections to sum rules.

Having obtained the NLO corrections to the coefficient functions we can take the case of $n = 1$ and evaluate the corrections to the sum rules of section 3.4. From table 5.2, we see that for $n = 1$, $B_{2,n}^{NS}$, $\gamma_1^{n,NS}$ and γ_0^n are all zero. From (5.62) we have $B_{3,1}^{NS} = -4$ and from (5.61) $B_{L,1}^{NS} = 8/3$ and since $B_{1,n}^{NS} = B_{2,n}^{NS} - B_{L,n}^{NS}$ then $B_{1,1}^{NS} = -8/3$. So for the $n = 1$ moments of F_1, F_2 and F_3 we get NLO corrections which are scheme independent,

$$\begin{array}{lll}
1 + B_{1,1}(\alpha_s/4\pi), & 1 + B_{2,1}(\alpha_s/4\pi), & 1 + B_{3,1}(\alpha_s/4\pi) \\
= 1 - \frac{2}{3}(\alpha_s/\pi), & = 1, & = 1 - (\alpha_s/\pi)
\end{array}$$
$$(5.75)$$

Thus there is no correction to the Adler sum rule (3.45); the Bjorken sum rule (3.46) becomes

$$\int_0^1 \mathrm{d}x(F_1^{\bar{\nu}p} - F_1^{\nu p}) = 1 - \frac{2}{3}\frac{\alpha_s}{\pi} \qquad (5.76)$$

and the Gross–Llewellyn Smith sum rule (3.47) becomes

$$\int_0^1 \mathrm{d}x(F_3^{\nu p} + F_3^{\bar{\nu}p}) = 6\left(1 - \frac{\alpha_s}{\pi}\right) \qquad (5.77)$$

From (5.76) we see that the Adler and Bjorken sum rules do not agree beyond leading order and so the Callan-Gross relation (3.15) is no longer satisfied. The longitudinal structure function is not zero at NLO since $C_{L,n}^{NS}(1, \bar{g}^2) = B_{L,n}^{NS}(\alpha_s/4\pi)$ with $B_{L,n}^{NS}$ given by (5.61). In fact the moments of F_L^{NS} are simply related to those of F_2^{NS}. Because the anomalous dimension is the same we can write

$$\int_0^1 \mathrm{d}x\, x^{n-2}F_L^{NS}(x, Q^2) = B_{L,n}^{NS}\frac{\alpha_s(Q^2)}{4\pi}\int_0^1 \mathrm{d}x\, x^{n-2}F_2^{NS}(x, Q^2)$$

$$= \frac{4}{3}\frac{\alpha_s(Q^2)}{\pi}\frac{1}{(n+1)}\int_0^1 \mathrm{d}x\, x^{n-2}F_2^{NS}(x, Q^2)$$
$$(5.78)$$

Finally we can apply QCD corrections to polarised structure function sum rules. In (3.77) both the non-singlet quantities T_3

and T_8 have been shown (Kodaira *et al.* 1979*a, b*) to pick up a factor $1 - (\alpha_s/\pi)$. The Bjorken sum rule (3.78) therefore becomes

$$\int dx[g_1^p(x, Q^2) - g_1^n(x, Q^2)] = \frac{1}{6}\left|\frac{g_A}{g_V}\right|_{np}\left(1 - \frac{\alpha_s}{\pi}\right) \qquad (5.79)$$

5.3.3 NLO corrections to singlet moments

The NLO corrections to the anomalous dimensions $\gamma^{n,qq}$, $\gamma^{n,GG}$, $\gamma^{n,qG}$ and $\gamma^{n,Gq}$ have been computed together with the coefficient functions $\overline{C}_n^q(1, \bar{g}^2) = 1 + B_n^q(\alpha_s/\pi)$ and $\overline{C}_n^G(1, \bar{g}^2) = B_n^G(\alpha_s/\pi)$ (Floratos, Ross and Sachrajda 1979, Bardeen *et al.* 1978). The quark coefficient functions are the same as for non-singlet for F_1, F_2 and F_3, i.e. $B_{i,n}^q = B_{i,n}^{NS}$. For the gluon we have

$$B_{L,n}^G = T_2(F)\frac{16}{(n+1)(n+2)} \qquad (5.80)$$

independent of the renormalisation scheme, and

$$B_{2,n}^G = T_2(F)4\left\{\frac{4}{n+1} - \frac{4}{n+2} + \frac{1}{n^2} - \frac{n^2+n+2}{n(n+1)(n+2)}\right.$$
$$\left. \times \left[1 + \sum_{j=1}^{n}\frac{1}{j}\right]\right\} \qquad (5.81)$$

in the $\overline{MS}$ scheme.

The equality between $\gamma^{n,qq}$ and $\gamma^{n,NS}$ at leading order is not maintained at next order because of the flavour singlet $q-q$ mixing diagram of fig. 5.8, i.e. $\gamma_1^{n,qq} \neq \gamma_1^{n,NS}$.

Using the relation (5.80) for the coefficients $B_{L,n}$ we can extend (5.78) to the general relation for the longitudinal structure function

$$\int_0^1 dx\, x^{n-2}F_L(x, Q^2) = \frac{\alpha_s(Q^2)}{4\pi}\left\{\frac{4}{3(n+1)}\int_0^1 dx\, x^{n-2}F_2(x, Q^2)\right.$$
$$\left. + \frac{2c}{(n+1)(n+2)}\int_0^1 dx\, x^{n-2}\, xG(x, Q^2)\right\} \qquad (5.82)$$

where $c = \sum e_q^2$ for electroproduction and $c = 4$ for neutrino scattering.

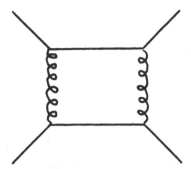

Fig. 5.8 Quark–quark mixing contribution which enters the calculation of $\gamma_1^{n,qq}$. Compact expressions for the $\gamma_1^{n,ij}$ can be found in Gonzalez-Arroyo, Lopez and Yndurain (1980).

At NLO the definition of the gluon distribution is not uniquely defined. We can expand the nth moment as

$$M_n^G(Q^2) = <O_n^G(Q^2)> $$
$$+ \frac{\alpha_s(Q^2)}{4\pi}\left[a_n <O_n^q(Q^2)> + b_n <O_n^G(Q^2)>\right] \quad (5.83)$$

where O_n^G, O_n^q are the singlet quark and gluon operators and a_n, b_n are arbitrary numbers. Expanding the singlet quark and gluon coefficient functions allows one to express the moments in the form

$$\begin{pmatrix} M_n^S(Q^2) \\ M_n^G(Q^2) \end{pmatrix} = \begin{pmatrix} 1 + B_n^q \frac{\alpha_s}{4\pi} & N_f B_n^G \frac{\alpha_s}{4\pi} \\ a_n \frac{\alpha_s}{4\pi} & 1 + b_n \frac{\alpha_s}{4\pi} \end{pmatrix}$$
$$\times \exp\left\{-\int_0^t dt' \gamma^{n,ij}(t')\right\}\begin{pmatrix} <O_n^q(Q^2)> \\ <O_n^G(Q^2)> \end{pmatrix} \quad (5.84)$$

We use the energy-momentum sum rule to fix up the ambiguity. The simplest choice to guarantee $M_2^S + M_2^G = 1$ is $a_n = -B_2^q$, $b_n = -N_f B_2^G$.

Finally we write down the generalisation of (5.15) to include the NLO corrections:

$$\begin{pmatrix} M_n^S(Q^2) \\ M_n^G(Q^2) \end{pmatrix} =$$
$$\begin{pmatrix} \frac{X_n(Q^2,Q_0^2)}{1-X_n(Q_0^2,Q^2)X_n(Q^2,Q_0^2)} & \frac{-Y_n(Q^2,Q_0^2)X_n(Q_0^2,Q^2)Y_n(Q^2,Q_0^2)}{Y_n(Q_0^2,Q^2)} \\ & \frac{-X_n(Q_0^2,Q^2)Y_n(Q^2,Q_0^2)}{Y_n(Q_0^2,Q^2)} \end{pmatrix}\begin{pmatrix} M_n^S(Q^2) \\ M_n^G(Q^2) \end{pmatrix}$$
$$(5.85)$$

where

$$X_n(Q^2, Q_0^2) = [X_n^{11} + X_n^{12}\alpha_s(Q^2) + X_n^{13}\alpha_s(Q_0^2)] \left(\frac{\alpha_s(Q^2)}{\alpha_s(Q_0^2)}\right)^{d_n^+}$$

$$+ [X_n^{21} + X_n^{22}\alpha_s(Q^2) + X_n^{23}\alpha_s(Q_0^2)] \left(\frac{\alpha_s(Q^2)}{\alpha_s(Q_0^2)}\right)^{d_n^-}$$

(5.86)

with a similar expression for $Y_n(Q^2, Q_0^2)$. The values of the X_n^{ij}, Y_n^{ij} in the $\overline{MS}$ scheme are listed by Floratos, Ross and Sachrajda (1979). Putting $Q^2 = Q_0^2$ gives the relations

$$X_n^{12} + X_n^{13} + X_n^{22} + X_n^{23} = 0 = Y_n^{12} + Y_n^{13} + Y_n^{22} + Y_n^{23} \quad (5.87)$$

Comparing with the leading order expression (5.15) we see that

$$X_n^{21} = a_n = 1 - X_n^{11}, \quad Y_n^{12} = b_n = -Y_n^{21} \qquad (5.88)$$

5.3.4 Corrections to sum rules
of polarised structure functions

We have already considered the QCD corrections to the non-singlet part of $g_1(x, Q^2)$ (see (5.79)) but when we turn to the singlet contribution, the issue is more complicated. Both contributions affect the integral of the polarised structure function $g_1(Q^2)$ (see (3.81)) which has been the focus of much attention. The QCD corrections to the singlet contribution (T_0 of (3.77)) are more subtle (Altarelli and Ross 1988) than the simple $(1 - \alpha_s/\pi)$ factor for the non-singlet contributions T_3 and T_8. The naive parton model result is not valid for T_0 due to axial anomaly (Adler 1969, Bell and Jackiw 1969). The only operator that contributes to T_0 is j_μ^5, where

$$j_\mu^5 = \sum_1^{N_f} \bar{q}_i \gamma_\mu \gamma_5 q_i \qquad (5.89)$$

Because of the axial anomaly, j_μ^5 is not conserved since

$$\partial_\mu j_\mu^5 = N_f \, \partial_\mu K_\mu \qquad (5.90)$$

where K_μ, the anomalous current, is

$$K_\mu = \frac{\alpha_s}{2\pi} \epsilon_{\mu\nu\rho\sigma} \, \text{Tr}[A^\nu (G^{\rho\sigma} - \tfrac{2}{3} A^\rho A^\sigma)] \qquad (5.91)$$

Although K_μ is gauge dependent, the diagonal matrix elements are indeed gauge invariant. The operator $j_\mu^5 - N_f K_\mu$ is therefore conserved and has vanishing matrix elements between one-gluon states. Because of the α_s in (5.91), we might expect this anomalous gluon contribution to T_0 to be suppressed but it is actually of the *same* order as the quark-parton contribution. For if we write

$$\Delta\Sigma(Q^2) = 9T_0 = \int_0^1 dx\, g_1^{singlet}(x, Q^2)$$

$$\Delta G(Q^2) = \int_0^1 dx\, [G^{(+)}(x, Q^2) - G^{(-)}(x, Q^2)]$$

(5.92)

then the evolution with Q^2 is given by (Altarelli 1982)

$$\Delta\Sigma(Q^2) = \Delta\Sigma(Q_0^2)$$

$$\Delta G(Q^2) = \Delta G(Q_0^2)\left[\frac{\alpha_s(Q_0^2)}{\alpha_s(Q^2)}\right] + \frac{4}{\beta_0}\Delta\Sigma(Q_0^2)\left[\frac{\alpha_s(Q_0^2)}{\alpha_s(Q^2)} - 1\right]$$

(5.93)

Thus the quantities $\Delta\Sigma(Q^2)$ and $\Delta\Gamma(Q^2) = (\alpha_s/2\pi)\Delta G(Q^2)$ are both constant (to leading order) with Q^2. The effect is therefore to replace the quark-parton quantity $\Delta\Sigma(Q^2)$ by the 'constituent-quark' quantity $\Delta\Sigma(Q^2) - N_f\Delta\Gamma(Q^2)$. Consequently, the experimental value of essentially zero (see discussion after (3.81)) attributed to $\Delta\Sigma$ in the quark-parton model could be interpreted as the value of $\Delta\Sigma - N_f\Delta\Gamma$. So the so-called 'missing spin' problem of the proton can be understood in terms of the parton model, together with an evolution of the quark densities by a sizeable anomalous gluon contribution. The QCD corrected form of the integral (3.77) is then

$$\int_0^1 dx\, g_1^p(x, Q^2) =$$
$$\tfrac{1}{12}\left[2S_z(u_+ - d_+) + \tfrac{1}{3}\cdot 2S_z(u_+ + d_+ - 2s_+)\right](1 - \alpha_s/\pi)$$
$$+ \tfrac{1}{9}\left[2S_z(u_+ + d_+ + s_+) - 3\Delta\Gamma\right](1 - [1 - 2N_f/\beta_0]\alpha_s/\pi)$$

(5.94)

The departure from the Ellis–Jaffe sum rule (3.80) can be attributed partly to a non-zero value of $\Delta s \sim -0.1$ and partly to a value $\Delta\Gamma \sim 0.13$ (Altarelli and Stirling 1989); for a full discussion see Ross (1989).

5.4 NLO corrections to the Altarelli-Parisi equations

5.4.1 Inverse Mellin transforms of moment evolution

Having obtained the NLO QCD correction to the evolution of the moments of structure functions, the easiest method to get the corresponding corrections to the structure functions themselves is to invert the definition of the moments. Take the non-singlet case,

$$M_n^{NS}(Q^2) = \int_0^1 dx\, x^{n-1} q^{NS}(x, Q^2)$$

$$\Rightarrow q^{NS}(x, Q^2) = \frac{-1}{2\pi i} \int_{a-i\infty}^{a+i\infty} dn\, x^{-n} M_n^{NS}(Q^2) \tag{5.95}$$

A similar relation holds between the anomalous dimensions and the splitting functions, e.g. for the leading order case we have already ((5.28), (5.34)) come across the relation

$$\gamma_0^{n,NS} = -4 \int_0^1 dz\, z^{n-1} P_{qq}(z)$$

$$\Rightarrow P_{qq}(z) = \frac{-1}{8\pi i} \int_{b-i\infty}^{b+i\infty} dn\, z^{-n} \gamma_0^{n,NS} \tag{5.96}$$

The values of a and b are chosen so that the contour of integration passes to the right of all singularities in the complex n plane. The extension of these relations to the NLO case was given by Herrod and Wada (1980) and Herrod, Wada and Webber (1981). For the qq splitting function we write

$$P_{qq}(z) = P_{qq}^{(0)}(z) + \frac{\alpha_s(Q^2)}{2\pi} P_{qq}^{(1)}(z) \tag{5.97}$$

where $P_{qq}^{(0)}$ is our previous leading order expression (5.34) and

$$P_{qq}^{(1)}(z) = \frac{-1}{16\pi i} \int dn\, x^{-n} \gamma_1^{n,qq} \tag{5.98}$$

The resulting parton distribution is not yet the physically observed structure function since we have not yet taken account of the NLO correction from the coefficient function. The analogue of (5.60) is

$$C^{NS}(x, \bar{g}^2) = \delta(1 - x) + B^{NS}(x) \frac{\alpha_s}{4\pi} \tag{5.99}$$

where $B^{NS}(x)$ is the inverse Mellin transform of B_n^{NS} which, for the case of F_2, gives

$$
B_2^{NS}(x) = C_2(F)\left[-\frac{3}{2}\frac{1+x^2}{(1-x)_+} + \frac{1}{2}(9+5x) - 2\frac{1+x^2}{1-x}\ln x \right.
$$
$$
\left. + 2(1+x^2)\left[\frac{\ln(1-x)}{(1-x)}\right]_+ - (9 + \tfrac{2}{3}\pi^2)\,\delta(1-x) \right]
$$

$$(5.100)$$

in the $\overline{MS}$ scheme. The structure function is finally expressed as a convolution of $C^{NS}(x)$ and the parton distribution

$$
\frac{1}{x}F^{NS}(x,Q^2) = q^{NS}(x,Q^2) + \frac{\alpha_s(Q^2)}{4\pi}\int_x^1 \frac{dy}{y}B^{NS}(\frac{x}{y})q^{NS}(y,Q^2)
$$

$$(5.101)$$

The two steps, i.e. A-P equation for the parton distribution and the convolution (5.101), may be concatinated into a single step

$$
\frac{dF^{NS}(x,Q^2)}{d\ln Q^2} = \frac{\alpha_s(Q^2)}{2\pi}\int_x^1 \frac{dy}{y}\left\{ P^{(0)}(\frac{x}{y}) \right.
$$
$$
\left. + \frac{\alpha_s(Q^2)}{2\pi}\left[P^{(1)}\frac{x}{y} - \frac{\beta_0}{4}B^{NS}(\frac{x}{y}) \right] \right\} F^{NS}(y,Q^2)
$$

$$(5.102)$$

Remember that here $\alpha_s(Q^2)$ includes NLO corrections – see (5.56). Using this procedure, the quark singlet and gluon distributions can be calculated. The final A-P equations for the parton distributions are

$$
\frac{dq_i(x,Q^2)}{d\ln Q^2} = \frac{\alpha_s(Q^2)}{2\pi}\int_x^1 \frac{dz}{z}\left\{ \sum_j \left[P_{q_i q_j}(z)q_j(\frac{x}{z},Q^2) \right.\right.
$$
$$
\left.\left. + P_{q_i \bar{q}_j}(z)\bar{q}_j(\frac{x}{z},Q^2)\right] + P_{q_i G}(z)G(\frac{x}{z},Q^2) \right\}
$$

$$(5.103)$$

with analogous equations for $\bar{q}_i(x,Q^2), G(x,Q^2)$. Some of the simplifications which hold for the leading order splitting functions are not maintained at NLO. Because of the contribution from the graph of fig. 5.8, $P_{q_i \bar{q}_j}$ is no longer zero, for example. Having evolved all the parton distributions the physical structure function is finally given by

$$
F(x,Q^2) = \int_x^1 \frac{dz}{z}\left\{ \sum_i C_i(z,\bar{g}^2)q_i(\frac{x}{z},Q^2) + C_G(z,\bar{g}^2)G(\frac{x}{z},Q^2) \right\}
$$

$$(5.104)$$

where the gluon coefficient function is

$$C_G(x, \bar{g}^2) = \frac{\alpha_s(Q^2)}{4\pi} B^G(x) \qquad (5.105)$$

where, for the F_2 structure function, in the $\overline{MS}$ scheme

$$B_2^G(x) = 4T_2(F)N_f \left\{ (1 - 2x + 2x^2) \ln\left(\frac{1-x}{x}\right) + 8x(1 - x) - 1 \right\} \qquad (5.106)$$

The full details of the expressions for all the quantities appearing in this section are given by Herrod and Wada (1980) and are equivalent to the corrections to the A-P equations derived by Curci, Furmanski and Petronzio (1980) and Floratos, Kounnas and Lacaze (1981).

What is the effect of changing the renormalisation scheme? We saw in section 5.3 that such a change can be characterised by a change to the coupling α_s. That is

$$\alpha_s \longrightarrow \alpha'_s = \alpha_s(1 - \frac{\beta_0}{2\pi}\alpha_s \ln \kappa)$$
$$\Lambda \longrightarrow \Lambda' = \frac{1}{\kappa}\Lambda \qquad (5.107)$$

It is easy to see from (5.102) that the evolution of the structure function is preserved to $O(\alpha_s)$ provided that we keep the splitting functions unchanged but modify the coefficient functions according to

$$B(z) \longrightarrow B'(z) = B(z) - 4P^{(0)}(z)\ln \kappa \qquad (5.108)$$

In fact, this is what we expect from taking the inverse Mellin transform of (5.68) and using (5.96).

5.4.2   *Longitudinal structure function and σ_L/σ_T*

Taking the expressions for $B_{L,n}^q$ and $B_{L,n}^G$ given by (5.61) and (5.80) and computing the inverse Mellin transforms gives the simple expressions

$$B_L^q(x) = C_2(F)\, 4x$$
$$B_L^G(x) = T_2(F)\, 16x(1 - x) \qquad (5.109)$$

Thus (5.82) can be inverted to give

$$F_L(x, Q^2) = \frac{\alpha_s(Q^2)}{\pi} \left\{ \frac{4}{3} \int_x^1 \frac{dy}{y} (\frac{x}{y})^2 F_2(y, Q^2) \right.$$
$$\left. + 2c \int_x^1 \frac{dy}{y} (\frac{x}{y})^2 (1 - \frac{x}{y}) y G(y, Q^2) \right\} \quad (5.110)$$

where $c = \sum e_q^2$ for electroproduction, $c = 4$ for neutrino scattering. This is an important signal for NLO corrections of the theory. Equation (5.110) says that $R = \sigma_L/\sigma_T \sim \alpha_s$ instead of being identically zero. A strong variation with x is expected at low x from (5.110). For many years the data on F_L or R was meagre and the errors very large. Recently the CDHSW collaboration (Berge et al. 1989), for example, has made a precise measurement of F_L which agrees beautifully with the prediction of (5.110) – see fig. 5.9.

Even the *fourth* order corrections to $F_L(x, Q^2)$ have been calculated (Duke, Kimel and Sowell 1982) but, in contrast to the enormity of the calculation, the correction turns out to be almost negligible.

At HERA it will be particularly exciting to study F_L in the very small x region which will be open to experiment.

5.4.3 Heavy flavour thresholds

So far we have considered Q^2 to be much greater than any quark masses. When production of heavy quarks is allowed however, then we should worry about the problem of threshold behaviour. Let us consider just one heavy flavour for simplicity, e.g. charm. The simplest procedure is to choose a value of $Q^2 = Q_0^2$ below which we insist on no charm production and then to generate charm above this value, via the gluon, as in fig. 5.10.

Evolving quark distributions via the A-P equations, the procedure is to define non-singlet distributions for *each* flavour i,

$$f_i^{NS}(x, Q^2) = \frac{1}{N_f} \Sigma(x, Q^2) - [xq_i(x, Q^2) + x\bar{q}_i(x, Q^2)] \quad (5.111)$$

where $\Sigma(x, Q^2)$ is the pure singlet combination for N_f active flavours. The quantities $f_i^{NS}(x, Q^2)$ and $\Sigma(x, Q^2)$ all evolve in Q^2 independently of each other. That is, at Q_0^2 we have $f_c^N(x, Q_0^2) =$

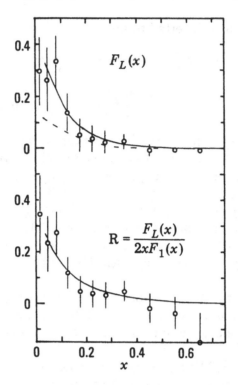

Fig. 5.9   Data on $F_L(x), R = \sigma_L/\sigma_T$ from the CDHSW collaboration (Berge *et al.* 1989). The curves are the result of using the F_2 and gluon distributions of Martin, Roberts and Stirling (1988*a*) in (5.110). The dashed curve shows the result of not including the gluon contribution.

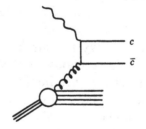

Fig. 5.10   Charm production via photon-gluon fusion.

$\frac{1}{N_f}\Sigma(x,Q^2)$ and at some higher Q^2 we recover the full charm distribution from (5.111) as $\frac{1}{N_f}\sum(x,Q^2) - f_c^{NS}(x,Q^2)$.

While this procedure is easy to carry out, it cannot be exact. For one thing, the threshold would appear more naturally in

$W^2 = Q^2(x^{-1}-1)$ rather than in Q^2. Another prescription (Glück, Hoffman and Reya 1982) uses the Bethe-Heitler description of the fusion process of fig. 5.10. The charm structure function (for $Q^2 \gg 4m_c^2$) is

$$F_2^C(x, Q^2) = 2e_c^2 \frac{\alpha_s(Q^2)}{2\pi} \int_{ax}^1 \frac{dy}{y} yG(y, Q^2) f(\frac{x}{y}, Q^2) \qquad (5.112)$$

where $a = 1 + 4m_c^2/Q^2$ and

$$f(z, Q^2) = v\left[4z^2(1-z) - \frac{z}{2} - \frac{2m_c^2}{Q^2}z^2(1-z)\right]$$
$$+ \left[\frac{z}{2} - z^2(1-z) + \frac{2m_c^2}{Q^2}z^2(1-3z) - \frac{4m_c^4}{Q^4}z^3\right] \ln\left(\frac{1+v}{1-v}\right)$$

with $v = 1 - [(4m_c^2/Q^2)z/(1-z)]$. Note that letting $m_c^2/Q^2 \to 0$ in (5.112) gives

$$F_2^c(x, Q^2) = e_c^2 \frac{\alpha_s(Q^2)}{4\pi} \int_x^1 \frac{dy}{y}(\frac{x}{y})\left\{B_{2,\overline{MS}}^G\left(\frac{x}{y}\right)\right.$$
$$\left. + 4P_{qG}^{(0)}\left(\frac{x}{y}\right)\ln\frac{Q^2}{m_c^2}\right\} yG(y, Q^2) \qquad (5.113)$$

where the gluon coefficient function (in the $\overline{MS}$ scheme) is given by (5.106). From (5.108) we see that (5.113) reproduces an effective change of renormalisation scheme, in this case a Q^2-dependent change. This hints towards incorporating the production of a heavy flavour by modifying the definition of $\alpha_s(Q^2)$ to include heavy quark masses. By taking the $\ln Q^2$ derivative of (5.112) we get the modified form of the P_{qG} splitting function for a heavy quark,

$$P_{cG}(z, Q^2) \equiv \frac{\partial f(z, q^2)/z}{\partial \ln Q^2}$$
$$= \frac{1}{v}\left[\frac{1}{2} - z(1-z) + \frac{m_c^2}{Q^2}\frac{z(3-4z)}{1-z} - 16\frac{m_c^4}{q^4}z^2\right]$$
$$- \left[2\frac{m_c^2}{Q^2}z(1-3z) - 8\frac{m_c^4}{Q^4}z^2\right] \ln\left(\frac{1+v}{1-v}\right)$$
$$(5.114)$$

and this can be used to evolve the heavy quark distribution in the Altarelli-Parisi equations.

Because β_0, β_1 depend upon the number of flavours, then for a given value of $\Lambda_{\overline{MS}}$ the running coupling $\alpha_s(Q^2)$ should reflect the

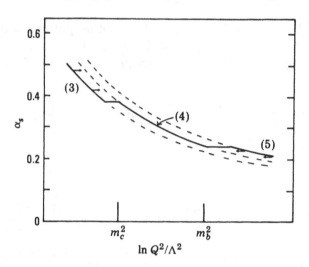

Fig. 5.11   Schematic representation of α_s for $N_f = 3, 4, 5$ showing how the three curves are modified into a single function by shifts along the Q^2 axis.

appropriate number of open flavours for each kinematic region. Fig. 5.11 shows α_s as a function of Q^2/Λ^2 for $N_f = 3, 4, 5$ together with the positions of the thresholds for charm and bottom production. One prescription for making a single $\alpha_s(Q^2)$ function across these thresholds is to shift horizontally two of the curves so as to match the values at $Q^2 = m_c^2$, $Q^2 = m_b^2$. This essentially is the prescription of Marciano (1984).

We start by defining $\alpha_s(Q^2, N_f)$ to be the solution of

$$\ln \frac{Q^2}{\Lambda_{(4)}^2} = \frac{4\pi}{\beta_0 \alpha_s} - \frac{\beta_1}{\beta_0^2} \ln \left[\frac{4\pi}{\beta_0 \alpha_s} + \frac{\beta_1}{\beta_0^2} \right] \qquad (5.115)$$

where $\beta_0 = 11 - \frac{2}{3} N_f$, $\beta_1 = 102 - \frac{38}{3} N_f$, i.e. we pick out $\Lambda_{\overline{MS}}$ corresponding to four flavours as our scale parameter. Thus we define

$$\alpha_{s(4)}(Q^2) = \alpha_s(Q^2, 4) \qquad (5.116)$$

and

$$\alpha_{s(5)}^{-1}(Q^2) = \alpha_s^{-1}(Q^2, 5) + \alpha_s^{-1}(m_b^2, 4) - \alpha_s^{-1}(m_b^2, 5) \qquad (5.117)$$

and

$$\alpha_{s(3)}^{-1}(Q^2) = \alpha_s^{-1}(Q^2, 3) + \alpha_s^{-1}(m_c^2, 4) - \alpha_s^{-1}(m_c^2, 3) \qquad (5.118)$$

Thus (5.117) ensures that $\alpha_{s(5)}(m_b^2) = \alpha_{s(4)}(m_b^2)$ and (5.118) that $\alpha_{s(3)}(m_c^2) = \alpha_{s(4)}(m_c^2)$, i.e. matching at thresholds. One could just as well take $4m_c^2, 4m_b^2$ as the thresholds, but the resulting change in α_s is extremely small. So for $Q^2 < m_c^2$ we take $\alpha_{s(3)}(Q^2)$ given by (5.118), for $m_c^2 < Q^2 < m_b^2$ we take $\alpha_{s(4)}(Q^2)$, and for $Q^2 > m_b^2$ we take $\alpha_{s(5)}$ given by (5.117). Again we could equally well have chosen to express α_s everywhere in terms of $\Lambda_{(5)}$ instead of $\Lambda_{(4)}$. We can express this matching procedure as an effective change of renormalisation scheme. For if we define $\alpha_{s(3)}(Q^2)$ to be given by (5.115) but with $\Lambda_{(4)} \to \Lambda_{(3)}$ then we get (Marciano 1984)

$$\Lambda_{(3)} = \Lambda_{(4)} \left(\frac{m_c}{\Lambda_{(4)}} \right)^{\frac{2}{27}} \left[\ln \left(\frac{m_c^2}{\Lambda_{(4)}^2} \right) \right]^{\frac{107}{2025}} \tag{5.119}$$

and

$$\Lambda_{(5)} = \Lambda_{(4)} \left(\frac{m_b}{\Lambda_{(4)}} \right)^{\frac{-2}{23}} \left[\ln \left(\frac{m_b^2}{\Lambda_{(4)}^2} \right) \right]^{\frac{-963}{13225}} \tag{5.120}$$

i.e. $\Lambda \to \Lambda' = \kappa^{-1}\Lambda$ which describes the effective change of scheme. If we took $\Lambda_{(4)}$ to be 100 MeV for example, (5.117), (5.118) give $\Lambda_{(3)} = 130$ MeV, $\Lambda_{(5)} = 63$ MeV.

The above prescription is practical and widely used (Collins and Tung 1986; Martin, Roberts and Stirling 1988a). One may criticise however that while $\alpha_s(Q^2)$ is continuous across the thresholds, the derivative is not. One can make even this continuous by using the Georgi-Politzer (1976) formula. The procedure follows from renormalising the *unphysical* quark mass so that the inverse propagator is $\not{p} - m$ at $p^2 = \mu^2$. Thus the quark mass m depends on μ^2 and the RGE is modified to take account of the β-function associated with the mass of each quark. For the charm quark we can write

$$\frac{M}{m_c} \frac{\partial m_c}{\partial \mu} = \beta_c(\bar{g}, \frac{m_c}{\mu}) \tag{5.121}$$

As a result the β function associated with the coupling is modified. To one-loop order the result is

$$\tilde{\alpha}_{s(3)}^{-1}(Q^2) - \alpha_{s(4)}^{-1}(Q^2) = \frac{1}{\pi} T_2(F) \mathcal{F} \left(\frac{m_c^2}{Q^2} \right) \tag{5.122}$$

where

$$\mathcal{F}(x) = -\frac{1}{3}\ln\frac{m_c^2}{M^2} + \frac{5}{9} - \frac{4}{3}x - \frac{1}{3}(1-2x)\sqrt{1+4x}\ln\frac{\sqrt{1+4x}+1}{\sqrt{1+4x}-1}$$

$$(5.123)$$

Here M^2 is the scale parameter appearing in (5.121). If we follow the momentum scheme procedure of putting $M^2 = Q^2$, (first term in (5.123) becomes $-\frac{1}{3}\ln x$) and then take the derivative of (5.122) w.r.t. $\ln Q^2$, we can express the the modified β-function in the form

$$\tilde{\beta}_0^{(3)} - \beta_0^{(4)} = -4T_2(F)x\frac{\mathrm{d}\mathcal{F}(x)}{\mathrm{d}x} \qquad (5.124)$$

Thus $\tilde{\beta}_0^{(3)}$ is the one-loop β function which smoothly interpolates between $\beta_0^{(3)}$ and $\beta_0^{(4)}$,

$$\tilde{\beta}_0^{(3)} = 11 - \frac{2}{3}\left[4 - 6x\left(1 - \frac{2x}{\sqrt{1+4x}}\ln\frac{\sqrt{1+4x}+1}{\sqrt{1+4x}-1}\right)\right] \qquad (5.125)$$

So $\tilde{\beta}_0^{(3)} \rightarrow \beta_0^{(4)}$ as $x \rightarrow 0$ $(Q^2 \gg m_c^2)$ and $\rightarrow \beta_0^{(3)}$ as $x \rightarrow \infty$ $(Q^2 \ll m_c^2)$. This type of continuous behaviour has been extended to the two-loop β-function, β_1 (Yoshino and Hagiwara 1984).

6

Large and small x

6.1 Large x: dimensional counting rules

In chapter 3 we mentioned the so-called counting rules which attempt to describe the shape of structure functions at large x. Given the Q^2 dependence described in chapters 4 and 5 the shape will change with Q^2 but we may think of some canonical scale μ^2 at which the exponents are supposed to apply. The counting rules were derived by Brodsky and Farrar (1973, 1975) and Matveev, Murddyan and Tavkheldize (1973) and extended by Farrar and Jackson (1975), Vainshtein and Zakharov (1978), Brodsky and Lepage (1980, 1981) and Chernyak and Zhitnitsky (1984). Excellent reviews are given by Close (1978) and Sivers (1982).

First consider the elastic form factors of a hadron; for a meson, the relevant graphs are shown in fig. 6.1.

In this graph the exchanged gluon carries momentum squared $\sim Q^2$, so the gluon propagator gives $1/Q^2$, the quark propagator gives $1/Q$ and the external quark lines altogether give a factor Q^2 So the net amplitude behaves $\sim 1/Q$, i.e.

$$< p_2|J_\mu|p_1 > \sim \text{const}/Q \sim (p_2 + p_1)_\mu \, \text{const}/Q^2 \qquad (6.1)$$

so the meson form factor $f(Q^2) \sim 1/Q^2$. In the case of a baryon, we get an extra gluon propagator $1/Q^2$, an extra quark propagator $1/Q$ and an extra pair of external quark lines giving factor Q and so get

$$< p_2|J_\mu|p_1 > \sim \text{const}/Q^3 \sim \bar{u}(p_2)\gamma_\mu u(p_1) \, \text{const}/Q^4 \qquad (6.2)$$

So, in general, we write the form factor of a hadron with n_H quarks as

$$f_H(Q^2) \sim \text{const}/(Q^2)^{n_H - 1} \qquad (6.3)$$

101

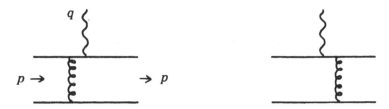

Fig. 6.1 One gluon exchange graphs for meson form factor.

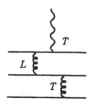

Fig. 6.2 Graph relevant for large x and transversely polarised photon.

But we have ignored the polarisation of the photon and we must ask whether (6.3) applies to transverse or longitudinal form factors. Because the quarks have spin $\frac{1}{2}$, the ratio of the coupling of a transversely polarised gluon to that of a longitudinally polarised one is $g_T/g_L = \sqrt{Q^2}$. Furthermore the graphs in fig. 6.1 correspond to Compton-like scattering where the dominant amplitude corresponds to one vector particle longitudinally polarised and the other transversely. So for the case of $n_H = 2$ we get $\sigma_L = Q^2\sigma_T$ where L, T refer to the photon polarisation. Adding another quark means there are two Compton scatterings, one transverse gluon and one longitudinal one. The dominant amplitude is for the transversely polarised photon and we get, in contrast to the meson case, $\sigma_T = Q^2\sigma_L$. So in using (6.3) we must remember that for systems with n_H even f_L is the leading form factor, while for those with n_H odd f_T is the leading one.

Turning to the large x behaviour of the structure functions and expressing the large x behaviour of the transverse structure function as

$$F_T(x) \sim (1-x)^k \sim (W^2/Q^2)^k \qquad (6.4)$$

the relevant graph (at $Q^2 \gg W^2$) is shown in fig. 6.2.

This is the graph which gave the $1/Q^3$ amplitude in the form factor case. Squaring this to get the structure function gives $k = 3$ in (6.4), i.e.

$$F_T^N(x) \sim (1 - x)^3 \qquad (6.5)$$

These arguments are only intuitive at best. A rigorous connection between taking the $x \to 1$ limit of the light-cone dominated DIS process and the large Q^2 behaviour of the form factor cannot really be given. Nevertheless applying (6.4) to the pion case, we saw that the transverse photon amplitude is suppressed $\sim 1/Q^4$ and so

$$F_T^\pi(x) \sim (1 - x)^2 \qquad (6.6)$$

For the longitudinal nucleon structure function the amplitude for the graph analogous to fig. 6.2 is also $1/Q^3$ but the structure function is suppressed by $1/Q^2$ from the longitudinal coupling of the photon, i.e.

$$F_L^N(x) \sim (1 - x)^3/Q^2 \qquad (6.7)$$

and the corresponding structure function for the pion is

$$F_L^\pi(x) \sim (1 - x)^0/Q^2$$

Thus the helicity structure of the QCD couplings is important and the general result for the large x behaviour of a quark of a given helicity in a hadron of a selected helicity state is given by

$$F_{q/H}(x) \sim (1 - x)^{2n_s - 1 + 2|\Delta h|} \qquad (6.8)$$

where n_s is the number of 'spectator' quarks in the hadron and Δh is the difference of the quark and hadron helicities. Thus, in a nucleon

$$\begin{aligned} F_{q^+/N^+}(x) &\sim (1 - x)^3 \\ F_{q^-/N^+}(x) &\sim (1 - x)^5 \end{aligned} \qquad (6.9)$$

To get the shape of the gluon distribution we convolute (6.9) with the radiation of a gluon from a quark and get (Close and Sivers 1979).

$$\begin{aligned} F_{G^+/N^+}(x) &\sim (1 - x)^4 \\ F_{G^-/N^+}(x) &\sim (1 - x)^6 \end{aligned} \qquad (6.10)$$

Going one step further and convoluting these distributions with a $G \to q\bar{q}$ distribution gives the antiquark behaviour at large x.

$$F_{\bar{q}^+/N^+}(x) \sim (1-x)^5$$
$$F_{\bar{q}^-/N^+}(x) \sim (1-x)^7 \tag{6.11}$$

These estimates (6.9)–(6.11) provide a useful guide to parametrising structure functions, for example at the starting value Q_0^2 of an analysis of DIS data using the Altarelli–Parisi equations. In particular it gives a first indication of the possible shape of the gluon distribution – at least for large x.

6.2   Large x: soft gluons beyond the renormalisation group.

Let us recall the evolution of the non-singlet structure function,

$$\frac{\mathrm{d}F(x,Q^2)}{\mathrm{d}\ln Q^2} = \frac{\alpha_s(Q^2)}{2\pi} \int_x^1 \mathrm{d}z P_{qq}^{(0)}(z) F(\frac{x}{z}, Q^2) \tag{6.12}$$

where the leading order splitting function is given by (5.34). Switching on the NLO corrections gives

$$\frac{\mathrm{d}F(x,Q^2)}{\mathrm{d}\ln Q^2} = \frac{\alpha_s(Q^2)}{2\pi} \int_x^1 \mathrm{d}z \Big\{ P_{qq}^{(0)}(z)$$
$$+ \frac{\alpha_s(Q^2)}{2\pi} \Big[P_{qq}^{(1)}(z) - \frac{\beta_0}{4} B^{NS}(z) \Big] \Big\} F(\frac{x}{z}, Q^2) \tag{6.13}$$

From (5.100) we see that $B^{NS}(z)$ contains a term proportional to $\ln(1-z)$. As $x \to 1$ we see that there will be correction terms $\sim \alpha_s(Q^2) \ln(1-z)$ which become increasingly important. At even higher orders we expect terms $\sim [\alpha_s(Q^2) \ln(1-z)]^k$ and the problem arises of summing all high order corrections of this type. In terms of moments, the correction terms have the form $[\alpha_s(Q^2) \ln^2 n]^k$ which can be large as n grows. The renormalisation group can cope with terms $\sim [\alpha_s(Q^2) \ln n]^k$ and resum them but it cannot resum terms with an extra power of $\ln n$. We are therefore interested in a region beyond the control of the renormalisation group.

The appearance of large logarithms like this is an indication that we are in a region where there is more than one relevant

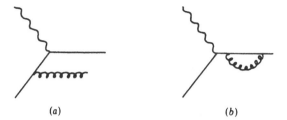

Fig. 6.3 The two types of gluon corrections (a) real gluon emission, (b) virtual gluon loops.

large invariant. For normal values of x, Q^2 and $W^2 = Q^2(\frac{1}{x} - 1)$ are both large with $Q^2 \sim W^2$, but as $x \to 1$ (with W^2 large) we have $Q^2 \gg W^2$ and these are our *two* large invariants. If we consider the corrections that arise from gluons, they are of two types - as shown in fig. 6.3.

At values x not close to 1, the singularities generated by real gluon emission and by virtual gluon loops exactly cancel, leaving the finite result. But as $x \to 1$ there is less and less phase space for the emission of the *real* gluons which become softer and softer. It is the mismatch between the singularities generated by real and virtual gluons which produces the $\ln^2 n$ terms.

The problem was overcome (Amati, Bassetto, Ciafaloni, Marchesini and Veneziano 1980, Brodsky and Lepage 1979) in an elegant way by simply replacing $\alpha_s(Q^2)$ in the *leading* order expression (6.12) by $\alpha_s[Q^2(1 - z)]$,

$$\frac{dF(x, Q^2)}{d\ln Q^2} = \frac{1}{2\pi} \int_x^1 dz \, \alpha_s[(1 - z)Q^2]P_{qq}^{(0)}(z)F(\frac{x}{z}, Q^2) \quad (6.14)$$

We can check that this picks up the correct $O(\alpha_s^2)$ correction by expanding α_s,

$$\frac{\alpha_s[(1 - z)Q^2]}{4\pi} = \frac{\alpha_s(Q^2)}{4\pi} + \ln(1 - z)\frac{d}{d\ln Q^2}\left(\frac{\alpha_s(Q^2)}{4\pi}\right) + \cdots$$

$$= \frac{\alpha_s(Q^2)}{4\pi} - \beta_0 \ln(1 - z)\left(\frac{\alpha_s(Q^2)}{4\pi}\right)^2 + \cdots$$

$$(6.15)$$

and substituting into (6.14) gives

$$
\frac{\mathrm{d}F(x,Q^2)}{\mathrm{d}\ln Q^2} = \frac{\alpha_s(Q^2)}{2\pi} \int_x^1 \mathrm{d}z \Big\{ P_{qq}^{(0)}(z)
$$
$$
+ \frac{\alpha_s(Q^2)}{2\pi} \left[-\frac{\beta_0}{4}\, 2\ln(1-z)P_{qq}^{(0)}(z) \right] \Big\} F(\frac{x}{z}, Q^2) \quad (6.16)
$$

By inspecting the $\ln(1-z)$ term in $B^{NS}(z)-(5.100)$, we see that (6.16) is consistent with the $O(\alpha_s^2)$ term expected from (6.13). If we had made a slightly different replacement, $\alpha_s(Q^2) \rightarrow \alpha_s(\widetilde{W}^2) = \alpha_s(Q^2(1-z)/z)$ then we would have also picked up the term $\sim \ln z$ in $B^{NS}(z)$. Actually Brodsky and Lepage (1979) argue that $\widetilde{W}^2$ is the relevant variable since it corresponds to the centre of mass energy of the virtual photon-parton scattering and this, in turn, specifies the maximum value of the transverse momentum which occurs in the ladder graphs which sum up to give the leading log result.

Using (6.14) to analyse DIS data poses some problems however. For fixed Q^2, as $z \rightarrow 1$, the argument of α_s becomes very small – indicating a breakdown of perturbative QCD. Furthermore, the argument of α_s, $\widetilde{W}^2$, is timelike and beyond leading order this implies that α_s becomes *complex*. This arises from the fact that higher order graphs involve time-like gluon momenta and we get terms $\sim \ln \widetilde{W}^2 = \ln |\widetilde{W}^2| - i\pi$ and the potentially large terms, proportional to π^2, appear at NLO.

A technique of analytic continuation from the space-like to the time-like region (Pennington and Ross, 1981) showed that the best replacement in (6.14) is $|\alpha_s(\widetilde{W}^2)|$, the modulus proving to be a much more efficient perturbative expansion paramter. Fig. 6.4 shows a comparison between the two forms for the expansion parameter.

We see that the variation of α_s is very much suppressed for small W^2/Λ^2. This behaviour could also be brought about by higher twist terms (see next section) by making the replacement $W^2/\Lambda^2 \rightarrow W^2/\Lambda^2 + C$, and since the z integration is dominated by $z \sim 1$ then it is expected to be sensitive to higher twist corrections.

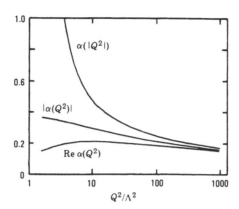

Fig. 6.4 Three expansion parameters, $\alpha_s(|Q^2|), |\alpha_s(Q^2)|$ and Re $\alpha_s(Q^2)$ in the timelike region.

6.3 Large x: higher twists

We have seen (section 4.3) that the leading singularities of the light cone expansion occur for those operators with the lowest twist, in this case twist $= 2$. Singularities from twist $= 4, 6, \ldots$ will be suppressed by powers of Q^2. At twist $= 2$ we had only two operators for any value of the spin n,

$$\bar{q}\gamma^{\mu_1} D^{\mu_2} \ldots D^{\mu_n} q \qquad \text{and} \quad G^{\mu_1 \nu} D^{\mu_2} \ldots D^{\mu_{n-1}} G^{\mu_n}_\nu \qquad (6.17)$$

but at twist $= 4$ there will be n operators with spin n of the type

$$\bar{q}\gamma^\mu D^{\mu_1} \ldots G_{\mu \mu_i} \ldots D^{\mu_n} q \qquad (6.18)$$

and n^2 operators of the type

$$\bar{q}\gamma^{\mu_1} D^{\mu_2} \ldots D^{\mu_i} \, q\bar{q} \, \gamma^{\mu_{i+1}} \ldots D^{\mu_n} q \qquad (6.19)$$

which correspond to the diagrams of fig. 6.5.

So, simply because there are at least n times as many operators contributing to the nth moment, we expect the Q^2 dependence of this moment to be modified typically as

$$M_n(Q^2) \longrightarrow M_n^{\tau=2}(Q^2)\left[1 + c\, n\, \frac{\Lambda^2}{Q^2}\right] \qquad (6.20)$$

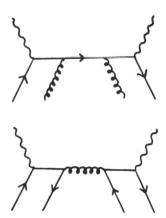

Fig. 6.5 Examples of graphs contributing to higher twist = 4 operators. (a) Two-quark process, (b) four-quark process.

and this would imply that the structure function is modified

$$F_2(x, Q^2) \longrightarrow F_2^{\tau=2}(x, Q^2) + \frac{\Lambda^2}{Q^2} \frac{d}{dx} F_2^{\tau=2}(x, Q^2)$$

$$\sim F_2^{\tau=2}(x, Q^2) \left[1 + \frac{\Lambda^2}{Q^2} \frac{c}{1-x} \right] \tag{6.21}$$

The phenomenological forms (6.20), (6.21) have been used in fits to data, attempting to estimate the size of possible higher twist contributions in an empirical way. Obviously one needs precise data at large x, small Q^2 and the analysis must take into account (known) target mass power corrections. A careful analysis along these lines has been carried out on low W^2 neutrino data (Varvell *et al.* 1987) and the results are shown in fig. 6.6. It is interesting that the higher twist term is estimated to be small and negative.

There has been considerable progress in the theoretical under-standing of higher twist terms following the proposal of Politzer (1980) that the proper procedure is to study whether such contri-butions can be expressed as the product of multiparton distribu-tions and corresponding parton cross-sections. If we formally take the Mellin transform of (4.46) we get

$$W(x, Q^2) = \sum_{\tau=2}^{\infty} \left(\frac{\Lambda^2}{Q^2} \right)^{\frac{\tau}{2} - 1} \int d\{y\} \, f_\tau(\{y\}, \mu^2) F_\tau(\{y\}, Q^2, \mu^2)$$

$$\tag{6.22}$$

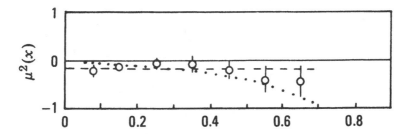

Fig. 6.6 Results of fitting x dependence of structure functions at low Q^2 by leading + higher twist components, $F(x, Q^2) = F^{\tau=2}[1 + \mu^2(x)/Q^2]$ (Varvell et al. 1987). Dashed line corresponds to $\mu^2(x) = -0.16$ GeV^2, dotted line to $\mu^2(x) = -0.41x/(1-x)$.

where $f_\tau(\{y\}, \mu^2)$ are parton correlation functions analogous to the single parton density distributions. Thus for twist $\tau = 2$ we have

$$\mathrm{d}\{y\} = \mathrm{d}y_1 \mathrm{d}y_2 \delta(y_1 y_2 - x)$$

$$\int \mathrm{d}y_1 \, y_1^n \, f_{\tau=2}(y_1, \mu^2) = \overline{O}_n^{\tau=2}(\mu^2)$$

$$\int \mathrm{d}y_2 \, y_2^n \, F_{\tau=2}(y_2, Q^2, \mu^2) = \overline{C}_{\tau=2n}(Q^2, \mu^2)$$

(6.23)

So by using the convolution formula (6.22) it is possible to make a parton model interpretation of higher twist contributions. At twist $\tau = 4$ (as at twist 2) the cross-sections $F_{\tau=4}(\{y\}, Q^2, \mu^2)$ can be computed in perturbative QCD but the correlation functions $f_{\tau=4}(\{y\}, \mu^2)$ are unknown. Things become complicated for $\tau = 4$ since several of these correlation functions appear in expressions for the structure functions. For example, for the two-quark operators (fig. 6.5a) we have two functions $\Lambda^2 T_-(x)$ and $\Lambda^2 T_+(x_1, x_2)$ – the single argument x arises when the graph is cut symmetrically and the $\pm$ refer to the relative helicity between the partons, i.e $\Lambda^2 T_-(x)$ is the correlation function of a quark and gluon with opposite helicities and total momentum fraction x.

The four quark operators (fig. 6.5b) give rise to two more correlation functions and the expression for the F_2 structure function is a complicated expression involving all four. However, it is possible to select physical quantities which pick out just one correlation function $\Lambda^2 T_-(x)$. Firstly the non-singlet combinations which enter the Gross–Llewellyn Smith and Bjorken sum rules involve only the integral of $\Lambda^2 T_-(x)$. Including the target mass corrections (section 5.1) and the NLO QCD corrections (5.3.2) we have (Ellis, Furmanski and Petronzio 1983, Miramontes 1985)

$$\int_0^1 dx \left\{ 1 - \frac{2}{3}\xi^2 \frac{M^2}{Q^2} - \frac{1}{3}\xi^4 \frac{M^4}{Q^4} \right\} \{ F_3^{\nu p}(x, Q^2) + F_3^{\bar{\nu} p}(x, Q^2) \}$$

$$= 6 \left[1 - \frac{\alpha_s(Q^2)}{\pi} - \frac{8}{3} \frac{\Lambda^2}{Q^2} < T_-^S(x) > \right] \quad (6.24)$$

and

$$\int_0^1 dx \left\{ 1 - \frac{2}{3}\xi^2 \frac{M^2}{Q^2} - \frac{1}{3}\xi^4 \frac{M^4}{Q^4} \right\} \{ F_1^{\bar{\nu} p}(x, Q^2) - F_1^{\nu p}(x, Q^2) \}$$

$$= 1 - \frac{2}{3}\frac{\alpha_s(Q^2)}{\pi} - 8\frac{\Lambda^2}{Q^2} < T_-^{NS}(x) > \quad (6.25)$$

where $< T_-(x) > = \int dx T_-(x)$. The labels S, NS refer to the flavour combinations $u_v + d_v$ and $u_v - d_v$ so that in the $SU(6)$ limit we have $< T_-^S(x) > = 3 < T_-^{NS}(x) >$. To estimate the magnitude of $< T_-(x) >$ some dynamics must be assumed. Bag model estimates (Iijima 1982, Fajfer and Oakes 1986) give $\Lambda^2 < T_-^{NS}(x) > \sim (80 \text{ MeV})^2$ and QCD sum rules (Braun and Kolesnichenko 1987) gives $(110$–$130 \text{ MeV})^2$ and this suggests that the twist-4 corrections to the sum rules (6.24), (6.25) are roughly equal to the NLO $\alpha_s(Q^2)$ corrections at $Q^2 = 1 \text{ GeV}^2$ and of the same sign.

The one structure function that involves only $T_-(x)$ is $F_L(x, Q^2)$. This is particularly interesting since the low Q^2 data from SLAC on F_L are not consistent with the leading twist expression (5.110). The full $1/Q^2$ corrections are (Miramontes 1985)

$$F_L(x, Q^2) = F_L^{\tau=2}(x, Q^2) + \frac{x^3 M^2}{Q^2}$$

$$\times \left[4\int_x^1 \frac{dy}{y^2} F_2(y, Q^2) - \frac{d}{dx}F_L^{\tau=2}(x, Q^2) \right] + 8\frac{\kappa^2}{Q^2}T_-(x) \quad (6.26)$$

which implies that $R = \sigma_L/\sigma_T$ is given by

$$R(x, Q^2) = R^{\tau=2}(x, Q^2) + \frac{4x^3 M^2}{Q^2} \frac{1}{F_2(x, Q^2)} \int_x^1 \frac{dy}{y} F_2(y, Q^2)$$

$$+ 8\frac{\kappa^2}{Q^2} \frac{T_-(x)}{F_2(x, Q^2)}$$

$$(6.27)$$

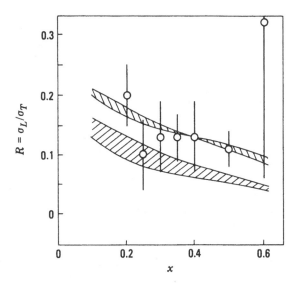

Fig. 6.7 Result of fitting $Q^2 = 5$ GeV^2 data on R from SLAC with (6.27). The lower dashed band is the leading twist result with NLO corrections and target mass corrections. The upper band is the result of adding the twist-4 contributions (Miramontes, Miramontes and Sanchez Guillen 1989).

where the second term is the target mass correction. Here κ is the unknown scale parameter for twist-4. Assuming soft gluon dominance (Ellis, Furmanski and Petronzio 1983), all the twist-4 matrix elements are equal and the twist-4 contribution $T_-(x)$ is proportional to $F_L^{\tau=2}(x)$ so that the last term in (6.27) is

$$F_L^{\tau=4}(x, Q^2) = 8\frac{\kappa^2}{Q^2}F_L^{\tau=2}(x, Q^2) \qquad (6.28)$$

which guarantees that

$$R(x, Q^2) \geq R^{\tau=2}(x, Q^2) \qquad (6.29)$$

which is in the required direction to explain the SLAC data, see fig. 6.7. In this case there is only *one* twist-4 operator (i.e. no weighting factor n for spin n) and so the higher twist contribution will *not* be associated with large x. Fig. 6.7 shows the result (Miramontes, Miramontes and Sanchez Guillen 1989) of fitting the SLAC data (Bodek *et al.* 1979, Dasu *et al.* 1988) with $\kappa^2 = 0.05$ GeV^2 which is consistent with the value of $\Lambda^2 T_-$ above. From the analysis of R it would seem clear that twist–4 terms are required in addition to the leading twist (NLO corrections included) plus consistent $O(M^2/Q^2)$ corrections. This procedure then avoids the problems associated with the ξ-scaling procedure.

6.4   Small x region: predictions of theory

Back in section 3.5 we saw that as $x \to 0$ we are exploring the high energy (high W^2) limit for the virtual photon–nucleon elastic forward scattering amplitude. The dominant exchange for this amplitude (fig. 3.3) is the Pomeron, and by studying the $x \to 0$ limit within the context of QCD we are really examining the QCD structure of the Pomeron.

To study the leading log prediction we take the A–P equations (5.39) for the evolution of the quark and gluon distributions. Rewriting the evolution equations in terms of s where

$$s = \ln \left[\frac{\ln Q^2/\Lambda^2}{\ln Q_0^2/\Lambda^2} \right] \tag{6.30}$$

we get

$$\frac{dG(x, s)}{ds} = \frac{2}{\beta_0} \int_x^1 \frac{dy}{y} \left\{ N_f P_{Gq}(\frac{x}{y})q(y, s) + P_{GG}(\frac{x}{y})G(y, s) \right\}$$

$$\frac{dq(x, s)}{ds} = \frac{2}{\beta_0} \int_x^1 \frac{dy}{y} \left\{ N_f P_{qq}(\frac{x}{y})q(y, s) + P_{qG}(\frac{x}{y})G(y, s) \right\}$$

$$\tag{6.31}$$

where we have assumed flavour independence of the quarks at small x. Now let $x \to 0$, $x/y \to 0$, then from section (5.1) we have

$$P_{Gq}(z) \longrightarrow C_2(F)\frac{1}{z} \qquad P_{GG}(z) \longrightarrow 2C_2(A)\frac{1}{z}$$

$$P_{qq}(z) \longrightarrow C_2(F) \qquad P_{qG}(z) \longrightarrow T_2(F)$$

$$\tag{6.32}$$

From (6.32), the gluon distribution is more singular than the quark distribution as $x \to 0$ and we need only keep the $G(y, s)$ on the r.h.s. of (6.31), giving a simple form of the evolution of the gluon distribution $xG(x, s)$:

$$\frac{d}{ds}xG(x, s) = \frac{1}{2}K \int_0^{\ln \frac{1}{x}} d(\ln \frac{1}{y})yG(y, s) \tag{6.33}$$

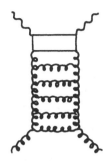

Fig. 6.8   Amplitude for virtual-photon–gluon scattering in the DLA.

where $K = 8C_2(A)/\beta_0$, so then

$$\frac{\mathrm{d}^2 xG(x,s)}{\mathrm{d}s\,\mathrm{d}\ln\frac{1}{x}} = \frac{1}{2}Kx\,G(x,s) \qquad (6.34)$$

For $\ln xG \gg 1$, (6.34) can be expressed as

$$\frac{\partial\ln xG(x,s)}{\partial s}\frac{\partial\ln xG(x,s)}{\partial\ln\frac{1}{x}} = \frac{1}{2}K \qquad (6.35)$$

which clearly has the solution

$$xG(x,s) = \exp\left[(2Ks\ln\frac{1}{x})^{\frac{1}{2}}\right] \qquad (6.36)$$

This rapid increase at small x is the result of taking the small x limit of the evolution equations in the double log approximation (DLA) where $\ln(1/x)$, $\ln\ln Q^2/Q_0^2 \gg 1$. The behaviour of the quark distribution at small x is dominated by the gluon ladder shown in fig. 6.8 and we can compute the quark distribution in the DLA to give

$$\begin{aligned}
xq(x,s) &= T_2(F)\sqrt{s/\beta_0 C_2(A)\ln\frac{1}{x}}\,xG(x,s)\\[4pt]
&= \frac{T_2(F)}{2C_2(A)}\frac{\partial xG(x,s)}{\partial\ln\frac{1}{x}}\\[4pt]
&= -\frac{x}{12}\frac{\partial xG(x,s)}{\partial x}
\end{aligned} \qquad (6.37)$$

The graph of fig. 6.8 is therefore the structure of the Pomeron in the DLA. The behaviour in (6.36), (6.37) agrees with the Regge asymptotic behaviour calculated in QCD by summing the leading $\alpha_s\ln(1/x)$ terms (Kuraev, Lipatov and Fadin 1977), the results coinciding in their common region of validity, $\ln Q^2/Q_0^2 \sim \ln(1/x) \gg 1$.

The important factor in deriving the DLA solution for the small x behaviour of the gluon given by (6.36) is the $1/z$ behaviour of the splitting function $P_{GG}(z)$ as $z \to 0$. This reflects the behaviour of the leading order contribution to anomalous dimension $\gamma_0^{n,GG}$ as $n \to 1$, namely $1/(n-1)$, see (5.13). In the limit $n \to 1$ it is actually possible to compute the anomalous dimension $\gamma^{n,GG}$ to all orders (Jaroszewicz 1980, 1982, Lipatov 1985)

$$\gamma^{n,GG} = -\frac{2C_2(A)\alpha_s}{\pi}\frac{1}{(n-1)}$$

$$\times \left[1 + \sum_{k=1}^{\infty} a_k \left(\frac{C_2(A)\alpha_s}{\pi}\right)^k \frac{1}{(n-1)^k}\right] \quad (6.38)$$

i.e. the higher order corrections are of the form $\alpha_s/(n-1)$. The corresponding form of the splitting function then becomes

$$P_{GG} = 2C_2(A)\frac{1}{z}\left[1 + \sum_{k=1}^{\infty} a_k \left(\frac{C_2(A)\alpha_s}{\pi}\right)^k \frac{(-\ln z)^k}{k!}\right] \quad (6.39)$$

Thus the pole in $\gamma^{n,GG}$ at $n = 1$ is replaced by a branch point singularity at $n = n_0$ where

$$n_0 = 1 + \frac{C_2(A)\alpha_s}{\pi}4\ln 2 \quad (6.40)$$

The gluon distribution is simply the inverse Mellin transform of its moments,

$$xG(x,Q^2) = -\frac{1}{2\pi i}\int_{a-i\infty}^{a+i\infty} dn\, e^{(n-1)\ln 1/x}\, G_n(Q^2) \quad (6.41)$$

where the moments are, in turn, expressed in terms of the anomalous dimension $\gamma^{n,GG}$,

$$G_n(Q^2) \sim \exp\left[-\int_0^t dt'\gamma^{n,GG}\right] \quad (6.42)$$

From (6.41) the behaviour of $xG(x,Q^2)$ will therefore be given by

$$xG(x,Q^2) \sim \exp[(n_0-1)\ln\frac{1}{x}]$$

$$\sim \frac{1}{x^D} \quad (6.43)$$

where $D = (n_0 - 1) = (C_2(A)\alpha_s/\pi)4\ln 2 \sim \frac{1}{2}$. Therefore the small x behaviour in the region where $\alpha_s \ln(1/x) \gg 1$ is even more dramatic than that of (6.36), i.e.

$$xG(x, Q^2) \sim \frac{1}{\sqrt{x}} \qquad (6.44)$$

The implications of such a singular behaviour for the gluon in hard processes at super collider energies have been pointed out by Collins (1985). This behaviour again reflects the nature of the Pomeron in the single gluon ladder approximation where $\alpha_s \ln \nu/M \gg 1$ (Balitzkii and Lipatov 1978). In fact n_0 given by (6.40) represents an *upper* bound of the Pomeron intercept; a *lower* bound can also be obtained, within this approximation, of $n_0 = 1 + 3.6\alpha_s/\pi$ (Collins and Kwiecinski 1988).

6.5 Very small x: parton–parton interactions

We saw in the previous section how the virtual-photon–photon cross-section, $\sigma^{\gamma^* p}$, violates the Froissart bound. In the region where $\alpha_s \ln(1/x) \gg 1$ the cross-section increased faster than $\ln \nu$. Clearly as x decreases further, new physics must enter to soften the asymptotic behaviour. The new ingredient turns out to be the self interactions of the gluons.

Let us go back to the gluon ladder of fig. 6.8 in the DLA. We may label each rung on the ladder by the fraction of energy carried, x_i, and by its transverse momentum q_{iT} – see fig. 6.9. The important region in the integration occurs when these variables are ordered:

$$x = x_n < \ldots < x_i < \ldots < 1$$
$$Q^2 \gg q_{nT}^2 \gg \ldots \gg q_{iT}^2 \gg \ldots \gg Q_0^2 \qquad (6.45)$$

We can use instead the pair of variables $\xi_i = \ln\ln(q_{iT}^2/\Lambda^2)$ and the rapidity $\eta_i = (C_A/\pi\beta_0)\ln(1/x_i)$ so that (6.45) becomes

$$\ln\ln\frac{Q_0^2}{\Lambda^2} = \xi_0 \ll \xi_1 \ll \ldots \ll \xi_i \ll \ldots \ll \xi_n \ll \xi = \ln\ln Q^2/\Lambda^2$$

$$0 < \eta_1 < \ldots < \eta_i < \ldots < \eta_n = \eta = \ln\frac{1}{x}$$
$$(6.46)$$

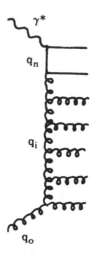

Fig. 6.9 Intermediate multigluon state in the DLA.

The average values of ξ_i and η_i are

$$< \xi_i > = \frac{i}{n}\xi \ , \quad < \eta_i > \simeq \frac{i}{n} < \eta > \qquad (6.47)$$

i.e. $< \eta_i >$ is proportional to $< \xi_i >$ and so ladder graphs can be represented by straight line trajectories in η, ξ plots. As Q^2 increases the ladder grows more rungs, the probability that x_n is 'large' *decreases* and number of partons with small x *increases*. As this number increases the probability of interactions between the partons naturally increases.

What determines the size of this interaction? A typical parton–parton cross-section, $\tilde{\sigma}$, is $\alpha_s(Q^2)/Q^2$ and ρ, the density of partons over the proton disk (of radius R_p), is $F(x, Q^2)/R_p^2$ and so the relevant parameter to characterise parton-parton interactions is ω, where

$$\omega = \rho\tilde{\sigma} = \frac{\alpha_s(Q^2)F(x, Q^2)}{Q^2 R_p^2} \sim \frac{\sigma^{\gamma^* p}}{R_p^2} \qquad (6.48)$$

For ω small, interaction between 'rungs' is negligible and so we have just graphs like fig. 6.9. From the last section we saw that $F(x, Q^2) \sim \exp\sqrt{2\xi\eta}$ in the DLA, so the condition for the single ladder approximation to be valid is $\omega \ll 1$ or

$$\eta \ll \frac{1}{\xi}\mathrm{e}^{2\xi} \qquad (6.49)$$

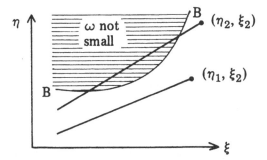

Fig. 6.10 η, ξ plot showing valid evolution (ξ_1, η_1) and invalid evolution (ξ_2, η_2) of single ladder graphs.

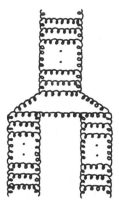

Fig. 6.11   Simplest example of multiladder graph representing gluon–gluon interaction.

This constraint is shown as the boundary B in fig. 6.10.

Evolution of ladder graphs corresponds to straight line trajectories in the plot of fig. 6.10. For example, evolution to the point (η_1, ξ_1) is valid because $\omega \ll 1$ everywhere along the line. Evolution to (η_2, ξ_2) is not straightforward since, for part of the trajectory, ω is sizeable and parton–parton interactions must be taken into account. These show up as multiladder graphs, an example of which is shown in fig 6.11.

If we associate the single ladder with the 'bare' Pomeron then the graph of fig. 6.11 may be regarded as a contribution to the triple Pomeron coupling. The result of including such contributions is to change the behaviour of the structure function at very small x and soften the increase coming from the bare Pomeron. These 'screening' corrections ensure that unitarity is not violated and the Froissart bound eventually satisfied. Gribov, Levin and Ryskin (1983) summed the graphs corresponding to the gluon-gluon interactions and specified the gluon density to depend on the range of x as follows. When x satisfies

$$\ln \frac{1}{x} < 0.21(\beta_0/4C_2(A)) \ln^2 \frac{Q_0^2}{\Lambda^2} \ln\ln \frac{Q^2}{\Lambda^2} \qquad (6.50)$$

the single gluon ladder approximation is valid, i.e. $xG(x, Q^2)$ is given by (6.36). When x satisfies

$$0.21 \frac{\beta_0}{4C_2(A)} \ln^2 \frac{Q_0^2}{\Lambda^2} \ln\ln \frac{Q^2}{\Lambda^2} < \ln \frac{1}{x} < 0.21 \frac{\beta_0}{8C_2(A)} \ln^2 \frac{Q^2}{\Lambda^2} \qquad (6.51)$$

then the gluon self-interactions modify the increase to give the approximation

$$xG(x, Q^2) \sim \exp\left\{ \left[2K \ln \frac{1}{x} \left(\ln\ln \frac{Q^2}{\Lambda^2} - \ln \sqrt{\ln \frac{1}{x}} + 1.1 \right) \right]^{\frac{1}{2}} \right\} \qquad (6.52)$$

For x values smaller than the range of (6.50) the graph-summation techniques break down and no precise behaviour is predicted.

Thus we see that if the gluon or parton density should get sufficiently large, i.e. if $\ln \frac{1}{x}$ and $\ln\ln Q^2$ are both very large, the self-interactions of the partons soften the small x behaviour to the form of (6.52). As far as deep inelastic scattering is concerned the modification is academic – the constraint (6.50) is easily satisfied even at HERA energies. The extremely small x values corresponding to (6.51), i.e. $x \sim 10^{-4}$, will be probed by hadronic collisions at the SSC; estimates (Kwiecinski 1985) are that the corrections to the gluon distributions are typically around 10%.

We saw that the softening of the small x behaviour is attributed to the increasing density ρ of gluons in the nucleon. Mueller and Qiu (1986) have shown how the Altarelli–Parisi equations are modified to take into account the effect of gluon recombination in the nucleon, in terms of a two-gluon correlation function $xG^{(2)}(x, Q^2)$. At small x we then have a non-linear evolution equation:

$$\frac{\partial x G(x, Q^2)}{\partial \ln Q^2} = \frac{\alpha_s(Q^2) C_2(A)}{\pi} \int_x^1 \frac{dy}{y} \, y G(y, Q^2)$$

$$-\frac{4\pi^3}{N_c^2 - 1} \left(\frac{\alpha_s(Q^2) C_2(A)}{\pi} \right)^2 \frac{1}{Q^2} \int_x^1 \frac{dy}{y} \, y^2 G^{(2)}(y, Q^2)$$

(6.53)

A reasonable guess for the correlation function might be $x G^{(2)}(x, Q^2) = [x G(x, Q^2)]^2 / \pi R_p^2$ and typically $x G(x, Q^2) \sim 3$ for small x so that the correction term is very small. On the other hand if we note the possible $1/\sqrt{x}$ behaviour of $x G(x, Q^2)$ (see section 6.4) then the correction at SSC could well be important. In the case of a nuclear target the gluon recombination may be more important, as we discuss in chapter 8.

7

The parton distributions

7.1 Analysing the DIS data

The accurate measurement of the structure functions $F_2(x, Q^2)$, $xF_3(x, Q^2)$, and $F_L(x, Q^2)$ means that we should be able to extract unpolarised parton distributions with a fair degree of precision from the deep inelastic data. The procedure consists of combining as much of the experimental data together, adjusting relative normalisations within the allowed systematic uncertainties and performing a QCD fit to the entire dataset. The QCD fitting procedure traditionally parametrises the parton distributions at some value of Q^2, $Q_0^2 = 4$ GeV^2 say, which is hopefully big enough that the unknown $O(\alpha_s^2)$ corrections are negligible, and evolving these distributions up in Q^2 with a particular value of $\Lambda_{\overline{MS}}$ via the Altarelli–Parisi equations. To minimise the effect of unknown higher twist contributions, the missing mass W^2 is cut at a value around $W^2 \sim 10$ GeV^2.

Ideally, the xF_3 neutrino data should be able to pin down the value of $\Lambda_{\overline{MS}}$ cleanly since the gluon distribution does not enter into the evolution. However the errors in xF_3 are significantly larger than those of F_2 since the former involves taking the *difference* of the ν and $\bar{\nu}$ cross-sections (see (3.35)). On the other hand a QCD analysis of the more precise data on F_2 involves mixing with the gluon whose distribution is, *a priori*, unknown. In general there is a correlation between the hardness of the gluon density and magnitude of Λ when analysing the F_2 data. For the leading order evolution of the flavour singlet F_2 involves two

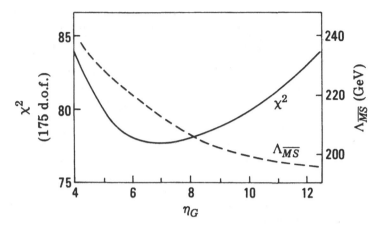

Fig. 7.1   Resulting χ^2 and $\Lambda_{\overline{MS}}$ as a function of η_G, where $xG(x, Q^2 = 4) \simeq (1 - x)^{\eta_G}$. NLO fit to the BCDMS data (Benvenuti *et al.* 1989) on $F_2^{\mu p}$.

terms,

$$\frac{dF_2(x, Q^2)}{d \ln Q^2} = \frac{\alpha_s(Q^2)}{2\pi} \int_x^1 dy$$
$$\times \left[P_{qq}(y) F_2(\frac{x}{y}, Q^2) + P_{qG}(y)(\frac{x}{y})G(\frac{x}{y}, Q^2) \right] \quad (7.1)$$

Taking $x \gtrsim 0.25$, the first term is negative and the experimental value of the derivative on the l.h.s. is negative. Thus if the second term was negligible we would extract some value of $\alpha_s(Q^2)$ from the data via (7.1). The second term is positive, and, as we increase the hardness of the gluon distribution, the effective magnitude of the first term is reduced and has to be compensated for by a larger value of $\alpha_s(Q^2)$ and consequently of Λ. Fig. 7.1 shows the result of fitting the precise data on $F_2^{\mu p}$ from the BCDMS collaboration (Benvenuti *et al.* 1989). One can see that the shape of the gluon distribution at low Q^2 is poorly constrained. In the final version of this experiment, the quoted systematic errors are substantially smaller than 2% so the resulting precision on η_G is, in principle, somewhat better.

In making a QCD fit to the combined data from different experiments, one should first be confident that the data is indeed consistent with the theory. This consistency can be tested by

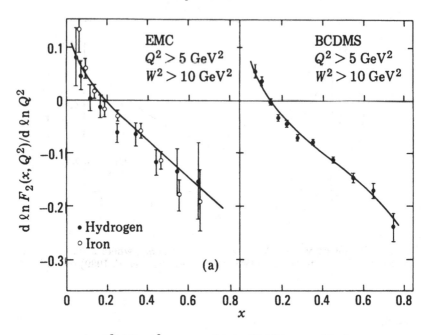

Fig. 7.2 $dF_2(x, Q^2)/d \ln Q^2$ versus x (a) for EMC iron and hydrogen targets, (b) BCDMS data.

comparing the $\ln Q^2$ derivatives of the various datasets with the corresponding quantities computed from the QCD fit. This is shown in fig. 7.2 where the data from the EMC (Aubert *et al.* 1985, 1986) and BCDMS collaborations are shown. Note that the EMC iron data are consistent with the theory, despite claims to the contrary (Voss 1987).

The fits we discuss in this chapter are the result of applying the Altarelli–Parisi equations with NLO corrections and using the prescription for heavy flavours generated by the gluon and described by (5.111) together with the threshold adjusted $\alpha_s(Q^2)$ described by (5.115)–(5.118). Thus our quoted values for Λ correspond to $\Lambda_{(4)}^{\overline{MS}}$. The results correspond to the analysis carried out by Martin, Roberts and Stirling (1988a). To cover a wide possibility for the gluon distribution at Q_0^2 one can parametrise the gluon as

$$xG(x, Q_0^2) = A_G x^\delta (1 - x)^{\eta_G} (1 + \gamma_G x) \qquad (7.2)$$

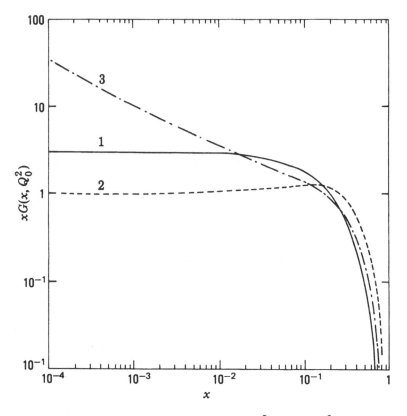

Fig. 7.3 The starting gluon distributions at $Q_0^2 = 4$ GeV^2 for the three parametrisations given in (7.3).

and choose three radically different shapes :

$$set\ 1: \delta = 0, \eta_G = 5, \gamma_G = 0,\ soft\ \text{gluon}\ (\Lambda_{\overline{MS}} = 100\ \text{MeV})$$
$$set\ 2: \delta = 0, \eta_G = 4, \gamma_G = 9,\ hard\ \text{gluon}\ (\Lambda_{\overline{MS}} = 250\ \text{MeV})$$
$$set\ 3: \delta = -\frac{1}{2}, \eta_G = 4, \gamma_G = 9,\ 1/\sqrt{x}\ \text{gluon}(\Lambda_{\overline{MS}} = 180\ \text{MeV})$$

$$(7.3)$$

The tail of the hard distribution is inspired by the standard parton distributions of Duke and Owens (1984) and the $1/\sqrt{x}$ gluon corresponds to the possibility of a singular behaviour discussed in section 4 of chapter 6. The three choices are shown in fig. 7.3.

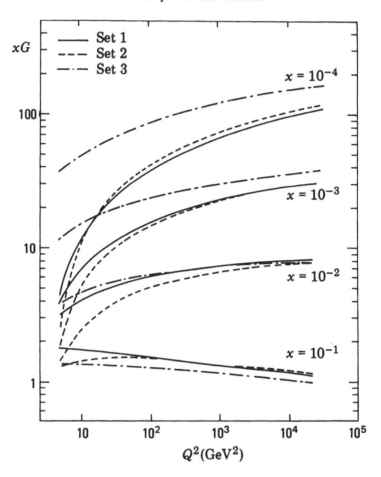

Fig. 7.4   The Q^2 dependence of the gluon distribution from each of the sets 1, 2, 3 at $x = 0.1, 0.01, 0.001$ and 0.0001.

7.2   The gluon and light quark distributions

Given a spread of gluon distributions at Q_0^2 it is interesting to see how the differences tend to fade away as Q^2 increases. Fig. 7.4 shows the Q^2 dependences of the three gluon distributions at small values of x and it is clear that the hard and soft gluons rapidly merge together. On the other hand the $1/\sqrt{x}$ gluon still differs significantly at very small values of x.

This figure shows how all gluon distributions develop a 'spike' at small x which is just the $\exp[\sqrt{(2Ks\ln 1/x)}]$ behaviour discussed

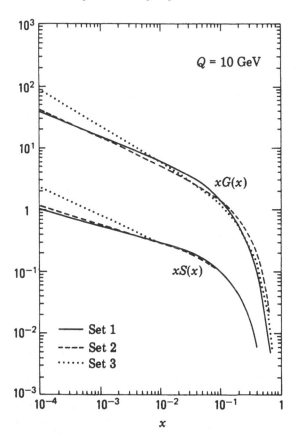

Fig. 7.5   The x-dependence of the gluon and the light quark sea $(\bar{u}, \bar{d}, \bar{s})$ at $Q^2 = 100$ GeV^2 of sets 1, 2, 3, equation (7.3).

in chapter 6.   This behaviour is mirrored in the sea quark distribution, shown in fig. 7.5, again as expected from (6.37). The valence distributions are shown in fig. 7.6 as a function of x, together with the light quark sea.

The distributions shown in figs. 7.4–7.6 are the results from fitting combined data from neutrino and muon experiments. There are two muon experiments, EMC and BCDMS, which do not agree. Fig. 7.7 shows the ratio of the $F_2^{\mu p}(x, Q^2)$ structure function measured by each experiment.

Both sets of data can be compared with the lower Q^2 data on F_2^{ep} from SLAC which has recently been completely re-analysed (Whitlow *et al.* 1989).   The conclusion is that the EMC data

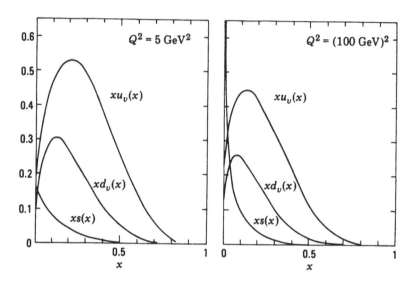

Fig. 7.6   The valence-quark and sea-quark distributions of set 1 at $Q^2 = 5$ and 10^4 GeV^2.

should be renormalised up by 8% and the BCDMS data bown by 2%, a procedure which removes the bulk of the discrepancy in fig. 7.7 at low x.

The most complete analysis to date (Harriman, Martin, Stirling and Roberts 1990) combines the DIS data ($F_2^{\mu p}$, F_2^n / F_2^p, $F_2^{\nu N}$) with the data on prompt photon production (see section 7.5) and the data on dilepton hadroproduction. The Drell-Yan cross-section at a centre of mass energy $\sqrt{s}$ takes the form

$$
\frac{\mathrm{d}^2 \sigma}{\mathrm{d}M\mathrm{d}y}\bigg|_{y=0} = \frac{16\pi\alpha^2}{9Ms} K \sum_q e_q^2 \, q(x, M^2) \, \bar{q}(x, M^2)
$$

$$
= \frac{16\pi\alpha^2}{9M^2\sqrt{s}} K \left[F_2^p(x, M^2) q_s(x, M^2) + q_s \times q_s \text{ terms} \right]
$$

$$(7.4)$$

where the sea is assumed to be $SU(3)$ symmetric ($\bar{u} = \bar{d} = \bar{s} = q_s$) and $x = M/\sqrt{s}$. Here M, y are the mass and rapidity of the dilepton pair. The factor $K(x, M^2)$ is determined by the exact $O(\alpha_s)$ NLO corrections (Altarelli, Ellis and Martinelli 1978, 1979). The analysis of Harriman *et al.* (1990) ends up with two distinct sets of parton distributions, one (HMRS(B)) consistent with the BCDMS data ($\times 0.98$) and the other (HMRS(E)) with the EMC

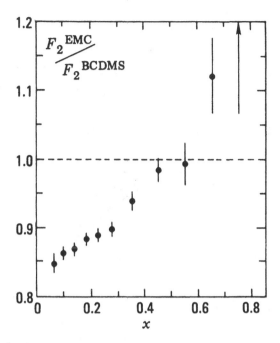

Fig. 7.7 Ratio of the structure function $F_2^{\mu p}$ measured by the BCDMS collaboration (Benvenuti *et al.* 1989) and by the EMC (Aubert *et al.* 1986).

data (×1.08). The former requires a value of $\Lambda_{\overline{M}S} \sim 200$ MeV with a gluon $\sim (1-x)^{5.1}$ and the latter requires a value of $\Lambda_{\overline{M}S} \sim 100$ MeV with a gluon $\sim (1-x)^{4.4}$. The latest Drell-Yan data from Fermilab (Brown *et al.* 1990) are very precise and fig. 7.8 shows the comparison of these data with to the fits of Harriman *et al.* This helps to pin down the sea-quark distribution to be $\sim (1-x)^{10}$.

For the moment we must therefore regard the parton distributions as having an uncertainty reflected by the ratios shown in fig. 7.9. and it is to HERA that we must turn for a resolution of this residual uncertainty.

7.3   The strange quark distribution

When single charm hadron production occurs in DIS followed by semi-muonic decay, the process is dominated by the diagram in fig. 7.10. The experimental signal for this is the observation of

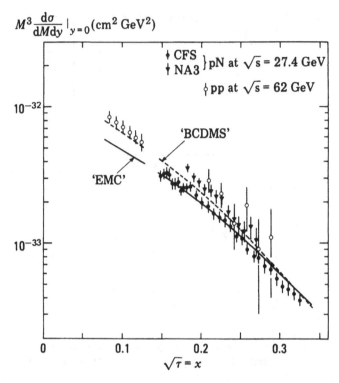

Fig. 7.8 Data on dimuon production from the E605 collaboration (Brown *et al.* 1990), in *pN* collisions at 800 GeV, together with the predictions of Harriman *et al.* (1990).

opposite-sign dimuons in the final state, one associated with the incoming ν or $\bar{\nu}$, the other from the c or $\bar{c}$ decay. From table 3.1 we see that the factors $f_i(x, \cos\theta_c)$ appearing in the cross-section for an isoscalar target are

$$\nu(p+n): f_i(x, \cos\theta_c) = x[2s(x)\cos^2\theta_c + (u(x) + d(x))\sin^2\theta_c]$$
$$\bar{\nu}(p+n): f_i(x, \cos\theta_c) = x[2\bar{s}(x)\cos^2\theta_c + (\bar{u}(x) + \bar{d}(x))\sin^2\theta_c]$$

(7.5)

Since $\sin^2\theta_c \simeq 0.053$ and $\bar{u}, \bar{d} \ll u, d$, about 90% of the $\mu^+\mu^-$ events in $\bar{\nu}N$ come from the term involving $x\bar{s}(x)$. The $W^\pm$ in fig. 7.10 interacts with a light quark to produce a massive quark and the usual procedure for accounting for the threshold suppression is the slow-rescaling prescription, described in section 3.6. Corrections must be made for $\mu^+\mu^-$ coming from muonic decays of $\pi's$ and $K's$ – about 6%.

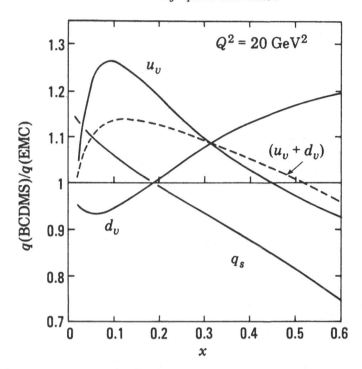

Fig. 7.9   Ratios of quark distributions at $Q^2 = 20$ GeV^2 obtained from global fits to the renormalised BCDMS and EMC μN data combined with neutrino data and prompt photon data and Drell-Yan data. (Taken from Harriman *et al.* 1990).

A high statistics experiment by the CCFR collaboration (Foudas *et al.* 1988) analysed 1529 $\mu^+\mu^-$ events in νN and 284 $\mu^+\mu^-$ events in $\bar{\nu} N$. Here the strange quark distribution was not constrained to be the same shape as the non-strange sea and was assumed to be of the form

$$s(x) \propto (1-x)^\alpha (\bar{u}(x) + \bar{d}(x)) \qquad (7.6)$$

From the analysis, $\alpha = 4.8 \pm 1.0$ indicating a very soft strange quark distribution and $\eta_s = 0.068 \pm 0.012$ where

$$\eta_s = \int_0^1 dx \, 2xs(x) \Big/ \int_0^1 dx \, (xu(x) + xd(x)) \qquad (7.7)$$

To extract the *size* of the strange quark distribution from (7.5) one more piece of information is required. By measuring the ratio of the integrals of $2xF_1(x, Q^2)$ and $xF_3(x, Q^2)$ the ratio

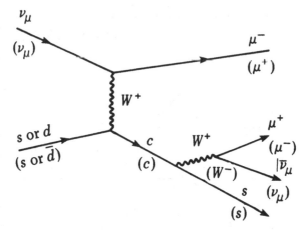

Fig. 7.10   Diagram for charm production in DIS giving rise to opposite-sign dimuon production.

$\bar{Q}/Q = \int \mathrm{d}x x q(x, Q^2)/\int \mathrm{d}x x \bar{q}(x, Q^2)$ can be estimated. Using $\bar{Q}/Q = 0.175 \pm 0.012$ we then get the ratio of the strange to non-strange sea to be

$$\kappa = \int_0^1 \mathrm{d}x \, 2 x s(x) \Big/ \int_0^1 \mathrm{d}x \, (x\bar{u}(x) + x\bar{d}(x)) = 0.46 \pm 0.11 \quad (7.8)$$

indicating that the sea is not $SU(3)$ symmetric ($\kappa = 1$). This analysis, although preliminary, suggests that the strange sea may only be roughly half that of the up or down sea. Presumably this suppression becomes less marked as Q^2 increases and the quark mass differences become less important.

7.4   The charm and bottom quark distributions

The heavy flavour quarks are generated from the gluon, as described in section 5.4. Taking threshold values of Q^2 to be 4 GeV^2 and 100 GeV^2 for charm and bottom production yields the distributions shown in fig. 7.11.

The charm structure function has been measured in muon production by the EMC (Aubert *et al.* 1983) and the BFP collaboration (Clark *et al.* 1980). Taking this simplest method of generating charm (i.e. using the evolution given by (5.111)) and varying the choice of the threshold Q^2 gives satisfactory agreement with the data, see fig. 7.12. Despite the apparent crudeness of the

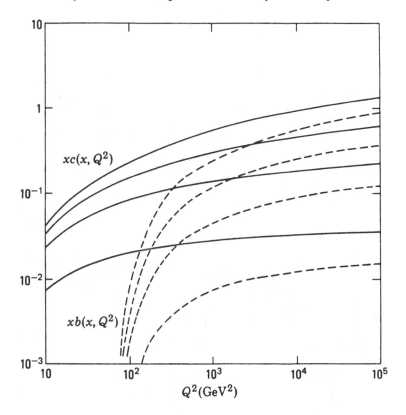

Fig. 7.11   Charm and bottom quark distributions versus Q^2 at fixed x values of $0.1, 0.01, 0.001$ and 0.0001.

threshold description, it is clear that ordinary QCD evolution, via the photon-gluon fusion, is consistent with charm production data.

7.5   Information on the gluon distribution from other processes

At the end of chapter 6 we saw that the small x region is where the gluon structure of the proton becomes the dominant factor. So any process at very high energies involving parton scattering will automatically be a critical probe of the gluon. From DIS at present day energies there is still considerable uncertainty on the gluon density function as we have seen by the wide choice of

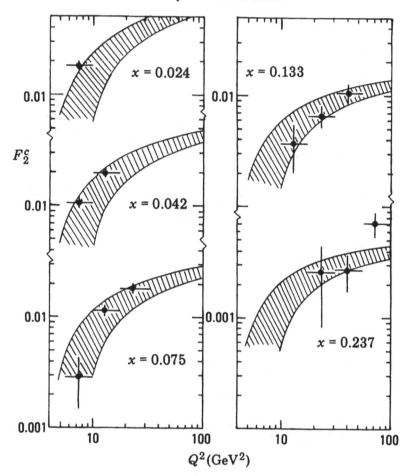

Fig. 7.12   Charm structure function measurements of EMC (Aubert *et al.* 1983) compared with the quantity $F_2^c(x, Q^2) = \frac{8}{9} c(x, Q^2)$ and $c(x, Q^2)$ is generated from zero at $Q_0^2 = 4, 8$ GeV^2 for the two curves shown. Taken from Martin, Roberts and Stirling (1989*b*).

possibilities for $xG(x, Q_0^2)$ which are consistent with the DIS data. Since any variation for the gluon density at low Q^2 values tends to diminish as Q^2 increases (see fig. 7.4) we may wonder, as we go to very high energies, why it matters what the precise gluon density is at low Q^2. The point is that for a process like jet production at collider energies ($\sqrt{s} = 630$ GeV) it is the transverse momentum or energy of the jet that is the relevant variable for setting the scale. As we probe small values of x ($x_T \sim 2p_T/\sqrt{s}$) then the

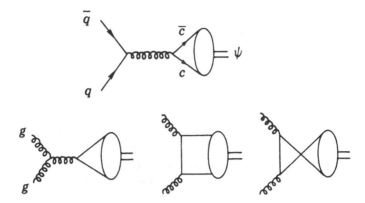

Fig. 7.13 Graphs relevant for ψ hadroproduction. In each case the emission of a soft gluon, to balance colour, is implied.

size of p_T will still be comparatively small. We now consider some examples of processes which are sensitive to the gluon distribution.

7.5.1 *Hadroproduction of* J/ψ

The collision of two protons to produce a J/ψ is supposed to be described by the graphs of fig. 7.13. There are several factors which make the calculation of the magnitude of the cross-section unreliable. One is the emission of soft gluons necessary for conserving colour and another is the wave-function which describes the probability for the produced $c\bar{c}$ pair to form either a ψ directly or a χ which subsequently decays into a ψ. The soft-gluon emission accounts for the p_T distributions of the ψ but the distribution functions of the partons in fig. 7.13 should describe the *form* of the longitudinal momentum distribution of the produced ψ. This distribution is usually described in terms of the Feynman x variable $\simeq 2p_L/\sqrt{s}$ where p_L is the longitudinal momentum. Neglecting the p_T of the ψ we can write

$$\frac{d\sigma}{dx_F} \propto x_1 G(x_1, Q^2)\, x_2 G(x_2, Q^2)$$

$$+ \frac{1}{r} \sum_q [x_1 \bar{q}_1(x, Q^2)\, x_2 q_2(x_2, Q^2) + (1 \longleftrightarrow 2)] \quad (7.9)$$

where $Q^2 \simeq M_\psi^2$, $r = \hat{\sigma}(GG \rightarrow \psi)/\hat{\sigma}(q\bar{q} \rightarrow \psi)$ and $x_{1,2} = \frac{1}{2}\left\{[x_F^2(1-\tau)^2 + 4\tau]^{\frac{1}{2}} \pm x_F(1-\tau)\right\}$ with $\tau = M_\psi^2/s$. The analysis

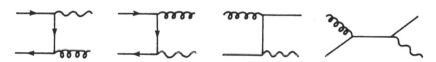

Fig. 7.14   Leading order graphs, $O(\alpha\alpha_s)$, for prompt photon production.

by Martin, Roberts and Stirling (1988a) using data on $\bar{p}N \to \psi$ and $pN \to \psi$ (to determine r) concluded that the data on the x_F distributions of the ψ tended to favour the parton distributions of set 1 over the other pair of sets. The steeply falling x_F distribution of the ψ likes to have a 'soft' gluon density.

7.5.2   Prompt photon production

The process $pN \to \gamma X$ provides a particularly clean example for appplying perturbative QCD. Firstly the cross-section, to leading order, is proportional to $\alpha\alpha_s$, thereby allowing a sensitive measure of α_s. This can be seen from the graphs of fig. 7.14.

Secondly the graphs of fig. 7.14 corresponding to the QCD Compton scattering $Gq \to \gamma q$ provide a way of testing the gluon distribution. Thirdly the photon provides an example of a jet in the final state where the problems of hadronisation are absent and so the experimental signature is clean. Finally it is a process where the NLO corrections have been computed and so the problem of the choice of scale, Q^2, for α_s and for the parton distributions has been solved. This was done by Aurenche and collaborators (Aurenche *et al.* 1984, Aurenche, Baier, Fontannaz and Schiff 1986, 1988) who also carried out an 'optimisation' procedure for choosing the values of the scales. Schematically the cross-section can be written

$$d\sigma = F(Q_1^2) \otimes F(Q_1^2)\alpha\alpha_s(Q_1^2)[d\hat{\sigma}^{(0)} + \alpha_s(Q_0^2) \otimes d\hat{\sigma}^{(1)}] \quad (7.10)$$

where $d\hat{\sigma}^{(0)}$ corresponds to the leading diagrams of fig. 7.14 and the higher order correction $d\sigma^{(1)}$ has an explicit dependence on the scales Q_0^2, Q_1^2

$$d\sigma^{(1)} = b\ln\frac{Q_0^2}{p_T^2}d\hat{\sigma}^{(0)} + 2 \otimes P_{qq}\ln\frac{p_T^2}{Q_1^2}d\hat{\sigma}^{(0)} + K \quad (7.11)$$

where K does not depend on Q_0^2, Q_1^2. Aurenche *et al.* take values for Q_0^2, Q_1^2 which minimise the sensitivity of the cross-section to the choice of the renormalisation scheme (with or without the constraint $Q_0^2 = Q_1^2$) i.e. $d\sigma/d\ln Q_i^2 = 0$ (Stevenson 1981a,1981b). Taking the single scale choice, $Q_0^2 = Q_1^2 \simeq 0.4(1 - x_T)p_T^2$ results in excellent agreement with the data, especially the precise data of the WA70 collaboration (Bonesini *et al.* 1987). Moreover the comparison with data suggests a soft gluon distribution at $Q^2 \simeq 5$ GeV^2. Taking the distribution to be $\sim (1 - x)^{\eta_G}$ gives $\eta_G = 4.3 \pm 0.3$ from the pp data for example. In contrast to the χ^2 distribution shown in fig. 7.1 for DIS data, the χ^2 distribution in this case is much narrower – emphasising the importance of prompt photon production to the gluon distribution. It is probably too naive to express the gluon density in terms of simply one parameter, η_G, and more important to realise which region of x is being probed by each process. In the case of the WA70 prompt photon data, the relevant x region for the gluon density is $x_T \sim 0.25$ to 0.6.

A related process is one where the jet associated with the quark or gluon in fig. 7.14 is detected. Martin, Roberts and Stirling (1988a) took the same choice of the optimised scale for $pp \rightarrow \gamma + \text{jet} + X$ as found for $pp \rightarrow \gamma + X$ and compared with data from the AFS collaboration (Akesson *et al.* 1987) assuming the three sets of parton distributions of (7.3). Again the preference is for the softer gluon of set 1.

7.5.3 Hadronic jet production

The process $p\bar{p} \rightarrow \text{jet} + X$ at collider energies is well understood in terms of hard parton scattering (Stirling 1987). The sub-processes include $qq \rightarrow qq, \bar{q}\bar{q} \rightarrow \bar{q}\bar{q}, q\bar{q} \rightarrow q\bar{q}, GG \rightarrow GG, GG \rightarrow q\bar{q}, Gq \rightarrow Gq, q\bar{q} \rightarrow GG$ and $G\bar{q} \rightarrow G\bar{q}$ so a significant part of the cross-section comes from gluon scattering. Moreover different regions of x_1, x_2 of the scattering partons are influenced by the different sub-processes. Fig. 7.15 shows how the small x_T region is dominated by the sub-processes involving two initial gluons.

Without the computation of the higher order corrections it is impossible to make an exact comparison with data on jet production since the choice of the scale is then arbitrary. However, choosing $Q = p_T/2$ does give an excellent description of the UA2 jet data, see fig. 7.16, if the parton distributions of set 1 are used. The parton distributions of sets 2 and 3 can give

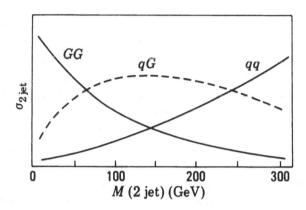

Fig. 7.15   Relative importance of sub-processes initiated by gluon-gluon, quark-quark and quark-gluon scattering for $\bar{p}p \rightarrow$ jet $+ X$ at $\sqrt{s} = 630$ GeV.

almost as good a description of these data provided the scale Q is varied to compensate for the change in magnitude for the cross-section caused by the change in Λ. Knowledge of the higher order corrections will allow the scale to be fixed and hopefully provide a discriminating test of the gluon distribution.

7.6   Pinning down the small-x gluon distribution

As we saw in the last section, hadronic production of jets should be able to provide a determination of the gluon distribution. The SSC, where very small x values will be probed, would therefore seem the ideal location for such a probe. In the meantime, HERA is well positioned to probe the gluon density for x down to 10^{-3}.

One way is to study inelastic lepto-production of J/ψ. The proposed mechanism for this process is photon-gluon fusion, shown in fig. 7.17.

The essential features of the model is that the J/ψ is formed at short distances, the spin-1 and colour singlet properties emerge directly from the hard scattering and long range effects like soft gluon emission can be neglected. The success of the mechanism can be judged only by comparison with experiment and the indication is that the description is adequate for sufficiently

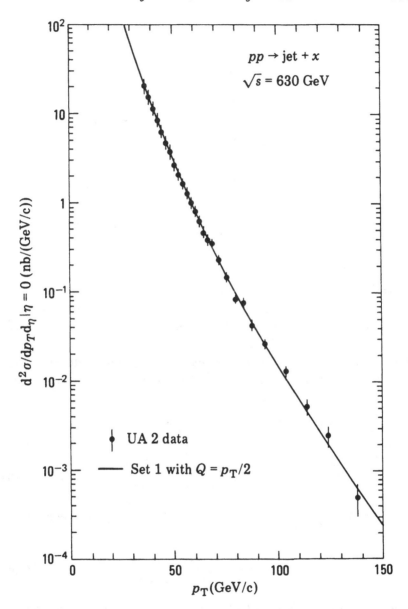

Fig. 7.16 The jet p_T distribution in $p\bar{p} \rightarrow$ jet $+ X$ at $\eta = 0$, $\sqrt{s} = 630$ GeV. Data are from UA2 (Appel *et al.* 1985). The curve corresponds to the choice of the soft gluon (set 1) with $Q = p_T/2$ (Martin, Roberts and Stirling 1988a).

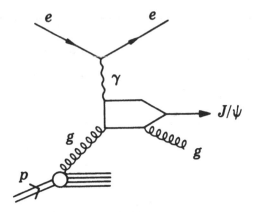

Fig. 7.17   Lowest order QCD diagram for $\gamma^* g \to J/\psi\, g$

inelastic $J/\psi's$. That is for

$$z = \frac{p_\psi \cdot p_N}{Q \cdot p_N} = \frac{E_\psi^{lab}}{E_{\gamma^*}^{lab}} < 0.8 \qquad \text{and} \qquad p_T^2/m_\psi^2 > 0.1 \qquad (7.12)$$

the model of fig. 7.17 can account for the muonproduction data if $\alpha_s(Q^2) \simeq 0.4$. The cross-section is governed by the gluon distribution,

$$\frac{d^2\sigma}{dz\,dp_T^2} = \frac{128\pi^2 G(x, \widehat{Q}^2) m_\psi \alpha_s^2 \alpha\, e_q^2 |R(0)|^2}{3s[m_\psi^2(1-z) + p_T^2]}$$
$$\times \left[\frac{1}{m_T^4} + \frac{(1-z)^4}{K_T^4} + \frac{z^4 p_T^4}{M_T^4 K_T^4}\right] \quad (7.13)$$

with $M_T^2 = m_\psi^2 + p_T^2$, $K_T^2 = m_\psi^2(1-z)^2 + p_T^2$, $\widehat{Q}^2$ is taken as m_ψ^2 and

$$x = \frac{1}{s}\left[m_\psi^2 z + \frac{p_T^2}{(1-z)z}\right] \qquad (7.14)$$

where $\sqrt{s}$ is the $\gamma^* N$ centre of mass energy. $R(0)$ is the radial wave function of the J/ψ at the origin, determined from the leptonic width.

From (7.13) we can write (Martin, Ng and Stirling 1987)

$$\frac{d\sigma}{dx} = xG(x, \widehat{Q}^2)\frac{\Gamma_{ee}}{m_\psi^3} f(x, s/m_\psi^2) \qquad (7.15)$$

where f is a known function, sharply peaked, just above $x_{min} = m_\psi^2/s$. As a result, the observed cross-section for inelastic J/ψ production is a direct measure of the gluon distribution for $x \simeq x_{peak}$. Thus by varying the virtual photon energy, ν, the x dependence of the gluon distribution can be determined (Tkaczyk, Stirling and Saxon 1988)

Another possibility is to extract the gluon distribution from a measurement of the longitudinal structure function at small x (Cooper-Sarkar *et al.* 1988). From (5.110) we see that F_L can be written, in the case of electroproduction, as

$$F_L(x, Q^2) = \frac{4}{3}\frac{\alpha_s(Q^2)}{\pi}\left[I_F(x, Q^2) + \frac{5}{3}I_G(x, Q^2)\right] \qquad (7.16)$$

where I_F, I_G are the contributions from the transverse and gluon structure functions, which can be expressed in the form

$$I_F(x, Q^2) = \int_0^{1-x} dz\,(1-z)F_2(\frac{x}{1-z}, Q^2)$$

$$I_G(x, Q^2) = \int_0^{1-x} dz\,z(1-z)\mathcal{G}(\frac{x}{1-z}, Q^2) \qquad (7.17)$$

where $\mathcal{G}(x, Q^2) = xG(x, Q^2)$. At small values of x, $F_2(x, Q^2)$ is dominated by the I_G contribution. Because x is very small we can make Taylor expansions about $x \simeq 0$ and since $F_2(x)$ is finite at $x \simeq 0$ we get $I_F \simeq \frac{1}{2}F_2(2x)$. For the gluon distribution we take $\tilde{\mathcal{G}}(x) = x^\delta G(x)$ to be finite at $x = 0$ with δ satisfying $0 \le \delta < 1$ although our prejudice is that the range is narrower, $0 \le \delta \lesssim \frac{1}{2}$ - see section (6.4). Expanding $\tilde{\mathcal{G}}(x)$ about $x = 0$ we get

$$I_G(x, Q^2) = \frac{1}{\theta}\mathcal{G}(\frac{1}{\xi}x) \qquad (7.18)$$

where

$$\xi = \frac{1+\delta}{3+\delta} \qquad \text{and} \qquad \theta = (3+\delta)(2+\delta)\xi^\delta \qquad (7.19)$$

To a good approximation we have $\theta \simeq 5.85, \xi = 0.4$ for a wide range of possible gluon distributions for $x < 0.1$. Because of the dominance of F_L by the gluon contribution of these small x-values inversion of (7.18) allows a precise determination of $xG(x, Q^2)$ at

HERA, i.e.

$$xG(x, Q^2) \simeq \frac{3}{5} \times 5.85 \left\{ \frac{3\pi}{4\alpha_s(Q^2)} F_L(0.4x, Q^2) - \frac{1}{2} F_2(0.8x, Q^2) \right\}$$
(7.20)

Therefore the gluon distribution at small x can be simply extracted provided that F_L can be measured at somewhat smaller values of x.

7.7   Polarised quark and gluon distributions

We define polarised quark and gluon distributions, $\Delta q(x, Q^2)$ and $\Delta G(x, Q^2)$ by

$$\Delta q(x, Q^2) = q_+(x, Q^2) - q_-(x, Q^2) + \bar{q}_+(x, Q^2) - \bar{q}_-(x, Q^2) \quad (7.21)$$

$$\Delta G(x, Q^2) = G_+(x, Q^2) - G_-(x, Q^2) \qquad (7.22)$$

where the $\pm$ refer to helicity $\pm\frac{1}{2}$ for the quarks and helicity ± 1 for the gluons. The structure function $g_1(x, Q^2)$ of the polarised proton can be expressed in terms of these distributions as

$$g_1(x, Q^2) = \frac{1}{2} \sum_q e_i^2 \left[\Delta q(x, Q^2) - \frac{\alpha_s}{2\pi} \Delta G(x, Q^2) \right] \qquad (7.23)$$

consistent with the contribution of the gluons to $\int dx g_1(x)$ (Altarelli and Ross 1988). Strictly, only this moment of $g_1(x, Q^2)$ is well defined but the form of (7.23) may be used to fix the higher moments.

From the latest estimate of $\int_0^1 dx g_1(x)$ (Ashman *et al.* 1989a) and the values of g_A/g_V from hyperon decays (see (3.75)), the contribution to each flavour at $Q^2 \simeq 10$ GeV^2 can be estimated:

$$\int_0^1 dx \, \Delta u(x, Q^2) \; - \Delta\Gamma \; = \; 0.782 \pm 0.032 \pm 0.046$$

$$\int_0^1 dx \, \Delta d(x, Q^2) \; - \Delta\Gamma \; = \; -0.472 \pm 0.032 \pm 0.046 \qquad (7.24)$$

$$\int_0^1 dx \, \Delta s(x, Q^2) \; - \Delta\Gamma \; = \; -0.190 \pm 0.032 \pm 0.046$$

where $\Delta\Gamma = (\alpha_s/2\pi) \int_0^1 dx \Delta G(x, Q^2)$. Adding the terms in (7.24) gives the result for the proton spin, $\Delta\Sigma - \Delta\Gamma = 0.060 \pm 0.047 \pm$

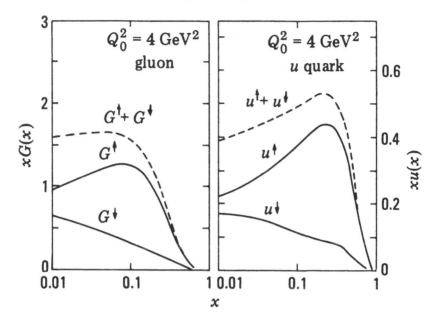

Fig. 7.18   Spin-dependent gluon and up-quark distributions estimated by Altarelli and Stirling (1989).

0.069. As discussed at end of section (5.3), $\Delta\Sigma$ and $\Delta\Gamma$ both remain essentially independent of Q^2. For an unpolarised sea, the strange quarks would satisfy $\int_0^1 dx\Delta s(x) = 0$ which, from (7.24), would require $\Delta\Sigma = 0.19$ and in this extreme situation we would have $\int_0^1 dx(\Delta u(x) + \Delta d(x)) = 0.7$. We can then write for the proton helicity

$$\frac{1}{2} = \frac{1}{2}\Delta\Sigma + \Delta G + L_z$$

$$= 0.35 + 6.3 - 6.15 \qquad (7.25)$$

where the large cancellation between the gluon helicity contribution and the orbital angular momentum contribution is consistent with QCD for $\Delta G + L_z$ is independent of Q^2 (Ratcliffe 1987).

Altarelli and Stirling (1989) extracted estimates for the individual polarised quark and gluon distributions assuming each to be of the form $\Delta q \sim N_q x^a (1 - x)^b$ at $Q_0^2 = 4$ GeV^2. By combining these with a parametrisation of the Q^2-dependent unpolarised distributions, they were able to make reasonable estimates of the parallel and anti-parallel spin components of the parton distributions. These are shown in fig. 7.18 where it is clear that a substantial part of the gluon distribution is polarised, comparable to that of the up-quark, in fact.

8

Quarks and gluons in nuclei – structure of the bound nucleon

8.1 Experimental situation

When we deeply inelastic scatter electrons, muons or neutrinos off a nuclear target we would, at first sight, expect the resulting structure functions to be little different from those measured off a hydrogen target. Essentially, it is the short distance scale that is being probed, i.e. as Q^2 increases we are resolving more and more of the quark and gluon structure of the nucleon. Putting a proton into a nucleus and subjecting it to the binding forces between nucleons would appear to affect only long-distance (i.e. ~ 1 fm) properties and so it is perhaps surprising to discover that there is a systematic modification to the structure function in a nucleus. The European Muon Collaboration (EMC) at CERN was the first to point out this modification which became known as the EMC effect (Aubert *et al.* 1982). Fig. 8.1 shows precise measurements of the effect over a wide range of Q^2. We summarise the experimental situation for $\rho = F_2^A(x, Q^2)/F_2^D(x, Q^2)$ as follows:

(a) As $x \to 1, \rho > 1$. This is to be expected. Both nuclear and nucleon structure functions are very small but the free nucleon structure function vanishes for $x > 1$ while the nuclear structure function vanishes strictly only for $x > A$.

(b) For $0.3 \lesssim x \lesssim 0.8$, $\rho < 1$.

(c) For $x = 0.25 \to 0.30$, $\rho \simeq 1$.

(d) For $0.1 \lesssim x \lesssim 0.25$, $\rho > 1$.

(e) For $x \lesssim 0.05$, ρ maybe < 1. This is nuclear shadowing.

(f) There is no evidence for any Q^2 dependence of ρ.

(g) There is a slow but quite definite variation of ρ with atomic mass, A.

In terms of parton distributions the clearest conclusion follows from (b) – valence quarks in a nucleus are degraded (have a lower $< x >$) relative to those in a free nucleon. Here we are assuming

143

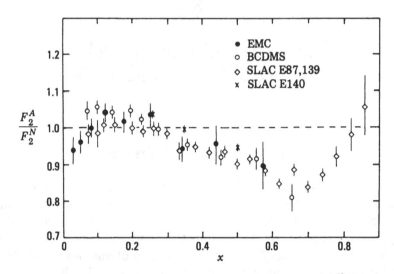

Fig. 8.1   Recent muonproduction data on the ratio of nuclear and deuterium structure functions. The EMC data (Arneodo *et al.* 1988, Ashman *et al.* 1988*b*,) uses a copper target, the BCDMS data (Voss 1987) an iron target. The average Q^2 varies with x, approximately $<Q^2> = (4 + 84x)$ GeV^2. The data from SLAC (Arnold *et al.* 1984, Dasu *et al.* 1988) has Q^2 varying from 3 to 15 GeV^2.

that a deuterium target is a good approximation to a free or unbound isoscalar nucleon. To make definite conclusions about sea-quarks and gluons is hard. There is no precise determination of the change in the integral of F_2 so we do not know directly if the fraction of momentum carried by the glue changes appreciably. The best probe of any modification to the sea quarks in a bound nucleon is the measurement of dilepton production by protons off a nuclear target. Here the cross-section is dominated by a beam valence quark (with x_1) annihilating with a target antiquark (with x_2). Fig. 8.2 shows the x_2 dependence of the ratio of antiquarks in the nucleus to those in deuterium and although these results are only preliminary there is evidence for only a small increase at low x.

8.2   The convolution model

It is natural to think of DIS off nuclear targets as a two-step process. The virtual photon scatters off quarks which are

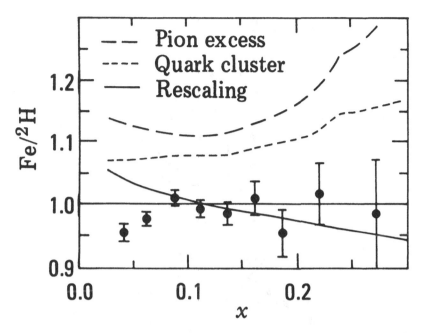

Fig. 8.2 Preliminary data from E772 at Fermilab (Brown *et al.* 1989) for hadronic dimuon production off iron and hydrogen targets.

distributed within nucleons which are, in turn, distributed within the nucleus. The two steps are indicated by the graph of fig. 8.3. The target nucleus A with momentum P has constituents T, momentum p and the virtual photon scatters off a quark a, of momentum k, from the constituent T (see Jaffe 1985). These constituents can, in principle, be states other than nucleons, e.g. pions, Δ's, 6-quark bags etc. The main assumption in writing the graph of fig. 8.3 is that only incoherent scattering off single target states is important. For example the graph of fig. 8.4 is ignored. Such a graph would be the analogue of the higher twist contribution of fig. 3.2 but in that case such contributions could safely be dropped as $\xi^2 \sim \frac{1}{Q^2} \to 0$ since they were suppressed by $\frac{1}{Q^2}$ factors.

For graphs like fig. 8.4 the distance scale η associated with the target distribution is not small in general and it is only under certain conditions (e.g. T= nucleon, with weak binding) that the neglect of such contributions can be justified.

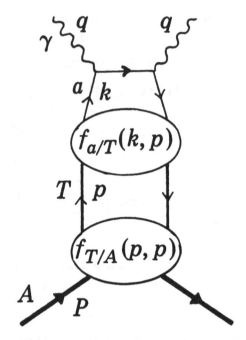

Fig. 8.3   Convolution framework for DIS off nuclear target A.

The form of the probability density of a quark in a nucleus, $f_{a/A}$ following from fig. 8.3 is

$$f_{a/A}(q^2, P \cdot q) = \sum_T \int \frac{d^4 p}{(2\pi)^4} f_{a/T}(p,q) f_{T/A}(P,p) \qquad (8.1)$$

The kinematics of DIS allow us to simplify this. Since

$$f_{a/T}(p,q) = f_{a/T}(p^2, p \cdot q, q^2) \text{ and } p \cdot q = p^+ q^- + p^- q^+ - \mathbf{p}_T \cdot \mathbf{q}_T$$

and $q^+/q^- \to 0$ then we can write for the distribution of quarks in constituent T,

$$f_{a/T}(p,q) \longrightarrow f_{a/T}(z = \frac{p^+ q^-}{q^2}, q^2, p^2) \qquad (8.2)$$

As for the distribution of constituent T with off-shell mass squared $\bar{p}^2$ in the nucleus A, we write

$$f_{T/A}(y_A, \bar{p}^2) = \int \frac{d^4 p}{(2\pi)^4} \, \delta \left(y_A - \frac{p^+}{P^+} \right) \delta(p^2 - \bar{p}^2) f_{T/A}(P,p) \quad (8.3)$$

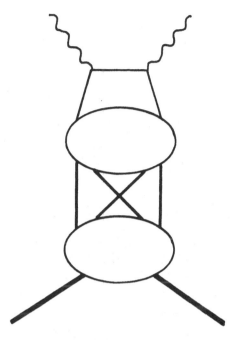

Fig. 8.4 Example of a non-convolution model graph.

and then the convolution form is just

$$f_{a/A}(q^2, P \cdot q)$$
$$= \sum_T \int \mathrm{d}\bar{p}^2 \mathrm{d}y_A \mathrm{d}z \delta(x_A - y_A z) f_{a/T}(z, q^2, \bar{p}^2) f_{T/A}(y_A, \bar{p}^2) \quad (8.4)$$

where $x_A = Q^2/2M_A\nu = k^+/p^+$. Note that $f_{T/A}(y_A, \bar{p}^2)$ is the probability to find constituent T in the nucleus A with momentum fraction y_A and invariant mass squared $\bar{p}^2$ while $f_{a/T}(z, q^2, \bar{p}^2)$ is the probability to find quark a with momentum fraction z in an (off-shell) target T with invariant mass $\bar{p}^2$. It is often *assumed* that $f_{a/T}$ does not depend on the scale $\bar{p}^2$ and that $\bar{p}^2 \simeq M_T^2$, M_T being the target on-shell mass. Thus we can finally express the convolution model in the form:

$$f_{a/A}(x_A, Q^2) = \sum_T \int \mathrm{d}y_A \, \mathrm{d}z \, \delta(yz - x_A) f_{a/T}(z, Q^2) f_{T/A}(y_A) \quad (8.5)$$

$$= \sum_T \int_{x_A}^1 \mathrm{d}y_A \, f_{a/T}\left(\frac{x_A}{y_A}\right) f_{T/A}(y_A) \quad (8.6)$$

Note that $0 \leq x_A \leq 1$ so that the Bjorken variable $x = \frac{M_A}{M_+} x_A$ satisfies $0 < x < A$, and that $0 < z = \frac{k^+}{p^+} < 1$ and $0 < y_A = \frac{p^+}{P^+} < 1$. The distributions are normalised by the $n = 1$ moments:

$$\int_0^1 \mathrm{d}z f_{a/T}(z) = N_{a/T} \qquad \int_0^1 \mathrm{d}y_A f_{T/A}(y_A) = N_{T/A} \qquad (8.7)$$

where $N_{T/A}$ is the number of constituents of type T in the nucleus A and $N_{a/T}$ is the number of quarks of flavour a in the constituent T. The $n = 2$ moments describe the fractions of momentum carried:

$$\int_0^1 \mathrm{d}z \, z f_{a/T}(z) = \zeta_{a/T} \qquad \int_0^1 \mathrm{d}y_A y_A f_{T/A}(y_A) = \zeta_{T/A} \qquad (8.8)$$

where $\zeta_{T/A}$ is the fraction of momentum P^+ of the nucleus carried by constituents T and $\zeta_{a/T}$ is the fraction of the constituent's momentum carried by quarks of flavour a. As a result of the convolution form, the moments are expressed in terms of products of moments; i.e. for any n,

$$M_n^{a/A} = \sum_T M_n^{a/T} M_n^{T/A} \qquad (8.9)$$

and so $N_a/A = \sum_T N_{a/T} N_{T/A}$ and $\zeta_{a/A} = \sum_T \zeta_{a/T}\zeta_{T/A}$. The quark distribution in the nucleus $F_{a/A}(x)$ is defined by $F_{a/A}(x) = \frac{M}{M_A} f_{a/A}(x_A)$ and we normally work with structure functions defined in terms of the quark distribution per nucleon, i.e. $\frac{1}{A} F_{a/A}(x)$.

It is important to note that we have described a very general formalism and not a specific model. Virtually no physics has gone into the model as yet (except for the neglecting of certain types of graphs like fig. 8.4). Progress is made only after we assume forms for $f_{a/T}$ and $f_{T/A}$. But the convolution model is a very useful framework for discussing almost all models of nuclear structure functions. These models differ basically in their different assumptions for the distributions $f_{a/T}$ and $f_{T/A}$. We shall discuss two approaches and show how they may be regarded as alternative descriptions of the same underlying physics.

The first thing to realise is that the concept of the constituent or 'virtual target' T is not well defined in the convolution framework (Close, Roberts and Ross 1988). We consider the graph of fig. 8.3 but for $T = N$, a nucleon; this is shown in fig. 8.5(a). If we look inside the top 'blob' we would find, some of the time, that the

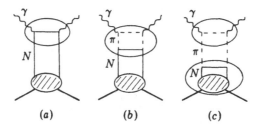

Fig. 8.5   (a) Convolution model graph for $T = N$. (b) View of (a) where $f_{a/N}$ contains virtual N + virtual π. (c) Virtual π in (b) now regarded as part of $f_{T/A}$ so that $T = \pi$.

the virtual photon is scattering off a pion target – as shown in fig. 8.5(b). We may consider this graph as essentially the same as fig. 8.5(c) which could be classified, in the convolution model, as choosing $T = \pi$. This ambiguity is not really surprising, it is related to a well known phenomenon we encountered in chapter 4 – the choice of a renormalisation point μ^2.

If we compare (8.6) with (4.49) the similarity of the convolution expressions for the quark densities is obvious.   In (4.49) the renormalisation scale μ^2 enters as a factorisation scale of which $F(x, Q^2)$ is independent. Varying μ^2 corresponds to changing the definition of the target states – in this case quarks and gluons. That corresponds, in fig. 4.5, to varying the definition of the lower 'blob'; i.e. how much of the gluon radiation from the quark a is included in the blob.   This is precisely the analogue of the ambiguity in the convolution model and we are free to specify definitions of the virtual targets T to suit our purpose. But there must be overall independence of the choice of the factorisation scale in the convolution model formalism – by this principle we can connect different versions of the model.

8.3   Conventional nuclear physics model

It is generally assumed that $f_{a/T}(z, q^2, \bar{p}^2)$ does not depend on the scale $\bar{p}^2$. This is the starting point of the conventional convolution model in which the modification of the nuclear quark distribution is attributed to the single-particle description of nucleons in a nucleus, i.e. the basic quark distributions in a bound nucleon are equal to those in a free nucleon. This corresponds to a particular

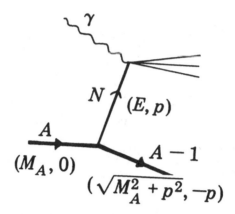

Fig. 8.6   Graph for DIS off a virtual nucleon. The $(A-1)$ nucleus is required to be on-mass-shell making the struck nucleon off-shell.

choice for the separation of the graphs in fig. 8.5 and implies that all the nuclear modification of the structure function is contained in $f_{T/A}(y_A)$.

Consider $T = N$ only. The momentum p of the virtual nucleon, in the lab frame, is

$$p = (M + \epsilon, \ \mathbf{p}) \tag{8.10}$$

where ϵ is the removal (or separation) energy given by $\epsilon = M_A - M_{A-1} - M$ and where the recoil kinetic energy of the nucleus has been neglected. The virtuality of the interacting nucleon, $\sqrt{p_0^2 - \mathbf{p}^2} - M$ is $\epsilon - T$ where $T = \mathbf{p}^2/2M$ is the kinetic energy of the nucleon. This nucleon is off-mass-shell simply because the $(A - 1)$ nucleus, in the intermediate state, has to be on-shell – see fig. 8.6. This is the requirement of the impulse approximation (Bickerstaff and Thomas 1987).

The assumption, in this approach, is that when the nucleon goes off-shell, the distribution of quarks is unaffected; i.e. $f_{a/N}(x, Q^2)$ is simply the distribution obtained from the structure function of the free nucleon. From (8.10) we get that

$$y_A = p^+/P^+ = [1 + (\epsilon + p_z)/M]/A \tag{8.11}$$

and we define $y = Ay_A$ so the average value of y is just

$$< y >= 1 + \bar{\epsilon}/M \tag{8.12}$$

$\bar{\epsilon}$ being the *average* removal energy of a nucleon. If there are only two-body binding forces in the nucleus then $\bar{\epsilon}$ and $\bar{T}$ may be related (Koltun 1972)

$$\bar{\epsilon} = T + V$$
$$2\mu = T + \bar{\epsilon} \tag{8.13}$$

where V is the binding potential (virtuality of the nucleon) and μ is the familiar binding energy (~ -8 MeV) . For a heavy nucleus, $T = \frac{3}{5}k_F^2/2M$ where k_F is the Fermi momentum of around 270 MeV, giving $T \simeq 23$ MeV and so from (8.13) we get $\bar{\epsilon} \simeq -40$ MeV.

Equation (8.12) thus describes the modification to the nucleon when it is bound in the nucleus, in terms of a single scale change. With this change, we can write $f_{N/A}(y_A) = \delta(y_A - <y_A>)$ and the convolution (8.6) gives simply

$$F_{a/A}(x, Q^2) \simeq F_{a/N}\left(\frac{x}{<y>}, Q^2\right) \tag{8.14}$$

where $<y>$ is given by (8.12). This is called x-rescaling (Akulinichev, Kulagin and Vagradov 1985). The point is that the off-shell nature of the struck nucleon introduces a scale change characterised by $\bar{\epsilon}$. It is, of course, too naive to take simply $f_{N/A}(y_A) = \frac{1}{A}\delta(y - z_0)$ where $z_0 = 1 + \bar{\epsilon}/M$; a more realistic choice would be a Fermi gas distribution,

$$f_{N/A}(y_A) = \frac{3mA^2}{4k_F}\left[1 - \frac{m^2 A^2}{k_F^2}(y_A - \frac{z_0}{A})^2\right] \tag{8.15}$$

for which the first few moments $M_n^{N/A}$ are

$$M_1^{N/A} = 1 \qquad M_2^{N/A} = z_0 \qquad M_3 = z_0^2 + k_F^2/5M^2 \tag{8.16}$$

From (8.9) we can now write the ratio of moments of the quark distribution in the nucleus to that in the free nucleon as

$$M_1^{q/A}/M_1^{q/N} = 1$$
$$M_2^{q/A}/M_2^{q/N} = z_0 \tag{8.17}$$
$$M_3^{q/A}/M_3^{q/N} = z_0^2 + k_F^2/5M^2$$

The term $k_F^2/5M^2$ is just $\frac{2}{3}T/M$. The first ratio says that the *number* of quarks is unchanged in going from a free to a bound nucleon; the second says that there has been a loss of the quark energy-momentum. It is this factor which gives the qualitative

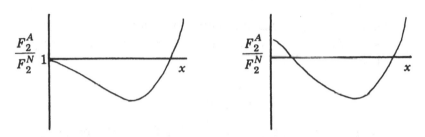

Fig. 8.7   Schematic display of the ratio of the nuclear and nucleon structure functions. (a) Taking only nucleons. The x-rescaling formula (8.14) is a good approximation at small x. The F.G. model (8.15) gives the description over all x. (b) Including the pion contribution modifies the ratio at low x.

description of the degrading of the valence quarks, approximately given by the x-rescaling of (8.14), $x \to x/z_0$. This is shown schematically in fig. 8.7(a).

To make up the missing energy-momentum, another virtual target T must be added to $T = N$. The natural candidate is the meson field – presumably that which is responsible for binding in the conventional nuclear physics description – and in particular the pion, $T = \pi$. In addition to the pion field associated with a free nucleon, there is supposed to be an *excess* of pions in the nucleus (Ericson and Thomas 1983, Llewellyn Smith 1983, Berger, Coester and Wiringa 1984). Thus there is a contribution to the nuclear structure function

$$\delta F_2^A(x, Q^2) = \int_x^1 \mathrm{d}y \, f_{\pi/A}(y) \, F_2^\pi(\frac{x}{y}, Q^2) \qquad (8.18)$$

where $F_2^\pi(x, Q^2)$ is the structure function of the pion and $f_{\pi/A}(y)$ is the distribution of excess pions in the nucleus. A reasonable description of $f_{\pi/A}(y)$ gives $< n_\pi > \sim 0.1 \to 0.2$ per nucleon, $f_{\pi/A}(y)$ tending to zero for large and small y, reaching a maximum around $m_\pi/m_N \simeq 0.15$. In fact the average value of y, $< y >$, must be $M_2^{\pi/A}/M_1^{\pi/A} = (1 - M_2^{N/A})/M_1^{\pi/A} = (1 - z_0)/ < n_\pi > \simeq 0.02/0.15 \simeq 0.13$. Thus the additional pion contribution (8.18) will be felt at low x values; the qualitative effect is shown in fig. 8.7(b).

The magnitudes of the quantities $\bar{\epsilon}$, T, V mentioned above led to rather good quantitative descriptions, and indeed such values appeared to follow from a harmonic oscillator model (Akulinichev, Shlomo, Kulagin and Vagradov 1985). However, the neglect of many-body forces in deriving (8.13) is very probably wrong (Li, Liu and Brown 1988) and consequently these values for $\bar{\epsilon}$ are too large. The derivation of the estimates for $\bar{\epsilon}$, V by Akulinichev *et al.* have been criticised (Frankfurt and Strikman 1987, 1988) and more realistic estimates give $\bar{\epsilon} \sim -26$ MeV. A detailed discussion of this issue in particular, and of the conventional nuclear physics model in general, is given by Bickerstaff and Thomas (1989).

The picture described in this section is self-consistent and may be thought of as a 'conventional' description in that the modification of the quark distribution is accounted for in terms of familiar properties of nuclear binding forces. Insofar that QCD is the theory that ultimately must account for the binding, not only of quarks and gluons, but between bound states of quarks and gluons, then the approach described above expresses the change to the nucleon structure function in terms of *long*-distance (~ 1 fm) QCD interactions.

8.4 Alternative convolution model

Having considered one 'extreme' form of the convolution model where the new environment of the nucleon was described in terms of a change to the long-distance physics which could be incorporated into $f_{T/A}$, we now consider the alternative extreme where $f_{T/A}$ is taken to be simply a delta function and we instead have a change of the short-distance physics which is described by $f_{a/N}$. This modified nucleon we shall label by N_A. So how does $f_{a/N_A}(z, Q^2)$ differ from $f_{a/N}(z, Q^2)$?

As in the conventional convolution model, the physics of the modification is characterised by a single scale change. In a heavy nucleus it is clear that individual nucleons are closely packed together: indeed it is very likely that neighbouring nucleons actually overlap. Consequently the confinement size of quarks (and gluons) is likely to be greater than in a free nucleon. So we may characterise the change by the confinement size, $\lambda \to \lambda_A$.

It is the confinement size which, in perturbative QCD, sets the scale for the low-momentum cut-off μ for radiating gluons, $\mu^2 \sim 1/\lambda^2$. Thus for a bound nucleon, N_A, we expect $\mu_A^2 < \mu^2$. When we evolve the quark distribution up in Q^2 then we suppose that the dominant change is due to this 'extra' evolution of the nuclear quark distributions. That is, assuming the starting distributions to be the same,

$$f_{a/N_A}(z, Q^2 = \mu_A^2) = f_{a/N}(z, Q^2 = \mu^2) \qquad (8.19)$$

then evolving to Q^2 in each case, gives a 'mismatch' of the structure functions

$$F_2^A(x, Q^2) = F_2^N(x, \xi_A Q^2) \qquad (8.20)$$

where

$$\xi_A \equiv \xi_A(Q^2) = \left[\frac{\lambda_A^2}{\lambda^2}\right]^{\alpha_s(\mu_A^2)/\alpha_s(Q^2)}$$

So the change of scale $\lambda \to \lambda_A$ leads to rescaling in Q^2 (Close, Roberts and Ross 1983). There is therefore an intimate connection between the scaling violations of the nucleon structure functions and the nuclear modification to the structure functions. If $\ln \xi_A$ is small we can expand the r.h.s. of (8.19) about Q^2 and get

$$\frac{F_2^A(x, Q^2)}{F_2^N(x, Q^2)} = 1 + \ln \xi_A(Q^2) \frac{\partial \ln F_2^N(x, Q^2)}{\partial \ln Q^2} + \cdots \qquad (8.21)$$

The derivative on the r.h.s. has the well-known behaviour shown in fig. 5.7 and so we expect the qualitative behaviour of the EMC effect as shown in fig. 8.8.

Good agreement with data is obtained by taking $\lambda_A = 1.15$ for $A = 56$, i.e. the confinement size of quarks increases by typically 15% in going to a heavy nucleus. In fact it is possible to construct a simple model of overlapping nucleons in the nucleus from which the A-dependence of ξ_A can be derived (Jaffe, Close, Roberts and Ross 1984) which accurately predicted the measured A-dependence of F_2^A (Close, Jaffe, Roberts and Ross 1985). Two features of the data not described by 'dynamic rescaling' are the drop at small x and the rise at large x. In the conventional nuclear physics framework the former is associated with the increase of the relevant longitudinal distances for photon nucleus scattering, usually termed 'shadowing' and the latter with the Fermi motion of the nucleons. In section 8.7 we shall indicate how both of these phenomena may be interpreted in a partonic way and indeed have a common origin, namely the recombination of partons. This $O(\alpha_s)$ correction is a modification of the short-distance behaviour

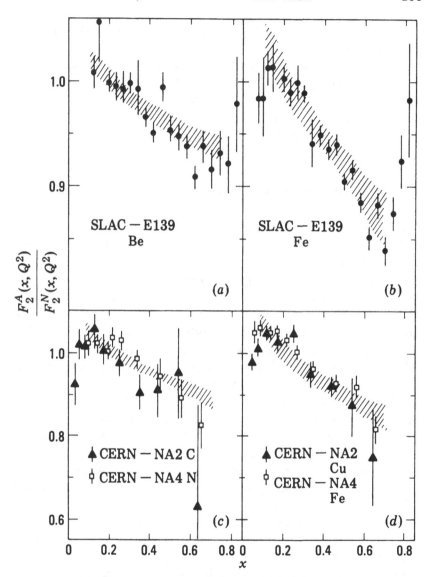

Fig. 8.8   Comparison of the data on F_2^A/F_2^N (Arnold *et al.* 1984, Bari *et al.* 1985, Benvenuti *et al.* 1987, Ashman *et al.* 1988b) with the r.h.s. of (8.21) shown by shaded bands. The uncertainties are due to extracting the logarithmic Q^2 derivatives directly from data.

of quarks and gluons and naturally fits into this version of the convolution model.

Putting such corrections aside for the moment, we deduce that the good agreement with data, seen in fig. 8.8, supports the notion that, in the language of the operator product expansion, the operator matrix elements of free and bound nucleons are related by a single scale change.

8.5   Relations between nuclear properties and QCD

We have now arrived at two versions of the convolution model, both of which give an adequate description of the bulk of the nuclear structure function data each being characterised by a single scale change. In section 8.2 we saw that different versions of the convolution model may be completely equivalent depending on the choice of $f_{a/T}$ and $f_{T/A}$ in the sense that the physical cross-section depends only on their product. This is essentially just the separation into short and long distance physics in the OPE where any particular choice corresponds to a particular value of the factorisation scale μ^2 – see (4.49).

Clearly the physical quantity $f_{a/A}$ cannot depend on the choice of factorisation scale, so if one choice corresponds to one version of the convolution model and another choice to another version then these two versions must be equivalent. Equating the two versions of sections 8.3, 8.4 will therefore give relations between the long distance nuclear physics properties and the short distance physics of QCD.

To derive the specific relations which follow from this idea of 'duality' between the two descriptions, it is convenient to consider moments of the distributions. For example (8.17) lists the A/N ratios of the $n = 1, 2, 3$ moments in the nuclear physical approach. In the dynamical rescaling approach we can write, for the non-singlet moments,

$$M_n^{q/A}(Q^2) = M_n^{q/N}(\xi_A Q^2) = \left[\frac{\alpha_s(\xi_A Q^2)}{\alpha_s(Q^2)} \right]^{d_n^{NS}} M_n^{q/N}(Q^2) \quad (8.22)$$

where $d_n^{NS} = \gamma_{0,n}^{NS}/2\beta_0$ and $\gamma_{0,n}^{NS}$ are the familiar non-singlet anomalous dimensions of QCD. The ratio of the α_s's appearing in (8.22) is independent of Q^2 and is given by

$$\frac{\alpha_s(\xi_A Q^2)}{\alpha_s(Q^2)} = \frac{\alpha_s(\mu^2)}{\alpha_s(\mu_A^2)} = 1 - \kappa_A \tag{8.23}$$

where μ^2, μ_A^2 are the scales appearing in (8.19) and κ_A is small ($\simeq 0.08$ for $A = 56$). We thus get, from (8.22)

$$\begin{aligned} M_1^{q/A}/M_1^{q/N} &= 1 \\ M_2^{q/A}/M_2^{q/N} &= (1 - \kappa_A)^{d_2^{NS}} \\ M_3^{q/A}/M_3^{q/N} &= (1 - \kappa_A)^{d_3^{NS}} \end{aligned} \tag{8.24}$$

and so equating (8.17) with (8.24) for $n = 2, 3$ gives

$$\begin{aligned} \bar{\epsilon}/M &= -d_2^{NS} \kappa_A \\ k_F^2/M^2 &= 5(2d_2^{NS} - d_3^{NS})\kappa_A + O(\epsilon^2/M^2) \end{aligned} \tag{8.25}$$

With $\kappa_A \simeq 0.08$, $d_2^{NS} = 0.427$, $d_3^{NS} = 0.667$ we get for $A = 56$, $\bar{\epsilon} \simeq -32$ MeV, $k_F \simeq 260$ MeV, neither of which are unreasonable. We can eliminate κ_A and use the equivalence of (8.17) and (8.24) to obtain a relation between the Fermi-momentum and mean removal energies:

$$\frac{k_F^2}{M^2} = 5\left(2 - \frac{d_3^{NS}}{d_2^{NS}}\right)\left(-\frac{\bar{\epsilon}}{M}\right) = \frac{35}{16}\left(-\frac{\bar{\epsilon}}{M}\right) \tag{8.26}$$

There are experimental data (Moniz et al. 1971) to support the proportionality of k_F^2/M^2 and $-\bar{\epsilon}/M$ with a constant of proportionality $\simeq 1.7$. From (8.26) this slope is predicted to be 2.2 but when $O(\alpha_s)$ corrections are included the value falls to around 1.75.

This illustrates how new results for old models can follow from the idea of factorisation scale independence in the convolution model of nuclear structure functions.

8.6 Polarised nuclei

We can similarly describe spin dependent structure functions of bound nucleons in terms of the convolution model. Recall that for a free nucleon we have the Bjorken sum-rule (3.72) which relates the integral of the difference of proton and neutron g_1 structure functions to the axial β-decay constant g_A/g_V. Thus, if we can relate the spin structure functions of a bound nucleon to those in a free nucleon (via rescaling) we can hope to relate g_A/g_V to G_A/G_V where the latter quantity is the axial constant for *nuclear* β-decay (Close, Roberts and Ross 1988).

In the convolution model we can express the distribution of quarks polarised parallel or anti-parallel to the target by

$$f_{a\uparrow/A} = f_{a\uparrow/N\uparrow}f_{N\uparrow/A} + f_{a\uparrow/N\downarrow}f_{N\downarrow/A} \qquad (8.27)$$

$$f_{a\downarrow/A} = f_{a\downarrow/N\uparrow}f_{N\uparrow/A} + f_{a\downarrow/N\downarrow}f_{N\downarrow/A} \qquad (8.28)$$

where we have used $f_{a\uparrow/N\uparrow} = f_{a\downarrow/N\downarrow}$ and $f_{a\uparrow/N\downarrow} = f_{a\downarrow/N\uparrow}$. The *unpolarised* quark distribution is obtained by adding (8.27) and (8.28) giving

$$f_{(a\uparrow+a\downarrow)/N} = [f_{a\uparrow/N\uparrow} + f_{a\downarrow/N\uparrow}][f_{N\uparrow/A} + f_{N\downarrow/A}]$$
$$= f_{a/N}f_{N/A} \qquad (8.29)$$

and in dynamical rescaling this becomes

$$f_{a/N_A}f_{N_A/A} = f_{a/N}(z, \xi_A Q^2)\, \delta(y_A - 1/A) \qquad (8.30)$$

Taking the *difference* of (8.27) and (8.28) gives the polarised quark distribution in the nucleus

$$f_{(a\uparrow-a\downarrow)/A} = [f_{a\uparrow/N\uparrow} - f_{a\downarrow/N\uparrow}] \cdot f_{N/A} \cdot a \qquad (8.31)$$

where the nuclear asymmetry, a, is

$$a = \frac{f_{N\uparrow/A} - f_{N\downarrow/A}}{f_{N\uparrow/A} + f_{N\downarrow/A}} \qquad (8.32)$$

To get a simple expression for a, consider conjugate nuclei, A and A^* (numbers of protons and neutrons reversed) with $J = \frac{1}{2}$. Since the nucleons have $J = \frac{1}{2}$ then a is simply

$$a = (S_o - S_e)/J \qquad (8.33)$$

where $S_{o,e}$ are the spins of the odd and even nucleons in the mirror nucleus. The nuclear shell model gives $S_e = 0$ and $S_o = \frac{1}{2}$ for $J = L_o + \frac{1}{2}$ and $S_o = -J/2(J+1)$ for $J = L_o - \frac{1}{2}$ which leads to $a = 1$ for $J = \frac{1}{2}^+$ and $a = -\frac{1}{3}$ for $J = \frac{1}{2}^-$.

We now take the $n = 1$ moment of (8.31) and use the Bjorken sum rule, to get

$$\left(\frac{G_A}{G_V}\right)^{A^*A} \bigg/ a = \left(\frac{g_A}{g_V}\right)_{\text{bd.nucleon}} \qquad \text{(independent of A)} \qquad (8.34)$$

Using values of 1 or $-\frac{1}{3}$ for a and the experimental values of G_A/G_V for several mirror nuclei ($A = 3 \rightarrow 31$) the r.h.s. of (8.34) varies by less than 20%.

It is possible to go a step further by relaxing the model dependent estimate of a given by (8.33). Buck and Perez (1983) related the magnetic moments μ_P, μ_N of mirror nuclei to $(S_o - S_e)/J$ and obtained a connection between μ_P, μ_N and $(G_A/G_V)^{A \cdot A}$ given by

$$\gamma_P - 1 = \frac{4.586}{\left(\frac{g_A}{g_V}\right)_{\text{bd.nucleon}}} \gamma_\beta + C_P$$

$$\gamma_N = \frac{-3.826}{\left(\frac{g_A}{g_V}\right)_{\text{bd.nucleon}}} \gamma_\beta + C_N$$

(8.35)

where $\gamma_{P,N} = \mu_{P,N}/J$ and $\gamma_\beta = (G_A/G_V)(S_o - S_e)/J$. Plotting the values of $\gamma_P - 1$ and γ_N against, Buck and Perez found a universal straight line. We can understand why the values lie on one line when we look at (8.34) and see that dynamical rescaling implies that $(g_A/g_V)_{\text{bd.nucleon}}$ is independent of A. The numerical value of this term suggested a 'quenching' of g_A/g_V for a bound nucleon (1.00 rather than 1.25 for a free nucleon) but even this discrepancy can be eliminated by including meson binding effects (Close, Roberts and Ross 1988).

8.7 Parton recombination

In considering DIS in a nucleus there is another feature associated with the close proximity (or even overlap) of neighbouring nucleons which affects the short distance physics of the partons. A parton associated with a particular nucleon can effectively 'leak' into a neighbour and fuse with a parton of that nucleon – this is shown schematically in fig. 8.9.

The physical consequences are felt most dramatically at large x and at small x. From fig. 8.9(a) it is clear that the quark receives a 'kick' from the gluon and if its momentum was already large then the quark which scatters with the virtual photon may then have a value of x which is greater than 1. Thus the nuclear structure function extends beyond $x = 1$ and the ratio $F_2^A(x)/F_2^N(x)$ will rise as x approaches 1. In fig. 8.9(d) two gluons fuse together to form a single gluon thereby reducing the effective gluon density,

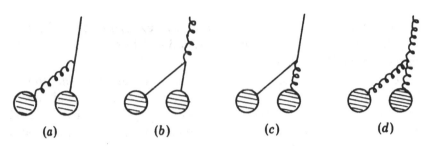

Fig. 8.9   Schematic view of parton fusion in the nucleus.

resulting in a sharp decrease of the gluon distribution in the nucleus at small x (shadowing of gluons). As we saw in section 6.4 the quark density as $x \to 0$ is dominated, as Q^2 increases, by the gluon density. Gluon shadowing becomes translated, therefore, into shadowing of the structure function at low x (Mueller and Qiu 1986).

Parton recombination is expected to be significant only when the actual spatial overlap between the fusing partons becomes sufficiently large, i.e. when the two-parton density in the nucleus is not too small. When does this occur for gluon fusion, for example? Consider a nucleus of radius R_A moving with high momentum: if each nucleon has momentum p, then the nucleus has a longitudinal size $\sim 2R_A(M/p)$. Gluons with a given x will have longitudinal size $\sim 1/xp$, so when $x < 1/(2MR_A)$ these gluons are forced to overlap with each other. At the same time these gluons are being probed by a photon with $q^2 = -Q^2$ and so they have a transverse size $\sim 1/Q$. As Q^2 increases the gluon density at very low x increases very rapidly (see sections 6.4, 6.5) and when it becomes much greater than $Q^2 R_A^2$ then the calculation of $xG(x, Q^2)$ itself becomes unreliable. So we shall consider only the situation when the two-parton density is not large, in order to apply leading logarithmic perturbative QCD.

The fusion of these partons, with transverse size $\sim 1/Q$ implies that they have transverse momentum $k_T^2 \sim Q^2$. The recombination cross-section goes like $1/k_T^2 \sim 1/Q^2$ and so one may expect that this higher twist behaviour would rapidly disappear as Q^2 increased. Mueller and Qiu (1986) showed that the Altarelli-Parisi equations are themselves modified by the recombination

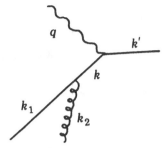

Fig. 8.10   Recombination of partons 1 and 2 to form parton 3.

of partons (see the extra term in (6.53)) with the result that the recombination cross-section, in particular the shadowing phenomenon, disappears extremely slowly as Q^2 grows (Qiu 1987). The observation of shadowing in the nuclear structure function (see fig. 8.1) shows no sign of going away at values of Q^2 as large as 7–10 GeV^2 and so gluon recombination is an attractive explanation for the suppression at small x and moderately large Q^2. Of course, as Q^2 decreases to very small values, ~ 0.1 GeV^2, the very marked shadowing seen there (Ashman *et al.* 1988*b*) is best explained in terms of generalised vector meson dominance of the virtual photon (Badelek and Kwiecinski 1988, 1989).

8.7.1 Parton fusion functions and the recombination contribution

The derivations of the contributions from parton recombination (Close, Roberts and Qiu 1989) follow from the same techniques (i.e. old-fashioned perturbation theory) used to derive the A-P equations in section 5.2. Consider the fusion of partons 1 and 2 to form parton 3, each with Bjorken x values x_1, x_2, x_3 as shown in fig. 8.10 than the modification $\Delta q(x)$ to the quark density $q(x)$ is

$$\Delta q(x) = \int dx_1 dx_2 \, T^{(2)}(x_1, x_2, \mathbf{k}_T^2) \frac{1}{\mathbf{k}_T^2} \Gamma_{p_1 p_2 \to p_3}(x_1, x_2, x_3) \delta(x_3 - x)$$

$$(8.36)$$

The momenta of the partons are given by

$$k_1 = (x_1 p + \frac{k_T^2}{2x_1 p}, \; k_T, \; x_1 p)$$

$$k_2 = (x_2 p + \frac{k_T^2}{2x_2 p}, \; k_T, \; x_2 p) \qquad (8.37)$$

$$k_3 = k = (x_3 p, \; 0, \; x_3 p)$$

The two partons, 1 and 2, are assumed to originate from *different* nucleons in the nucleus and already to have undergone QCD evolution before fusing to form parton 3. This means that $k_T^2 \sim Q^2$, giving a $1/Q^2$ factor to $\Delta q(x)$. The analogue of the splitting function in the A-P equations is now the parton-parton fusion function given by

$$\frac{1}{k_T^2} \Gamma_{1,2 \to 3}(x_1, x_2, x_3) = \frac{E_k}{E_{k_1} + E_{k_2}} |M(k_1 k_2 \to k)^2|$$

$$\times \frac{1}{(E_{k_1} + E_{k_2} - E_k)^2} \frac{1}{(2E_k)^2} \qquad (8.38)$$

The matrix element $M(k_1 k_2 \to k)$ is essentially the same quantity appearing in (5.30) and so $\Gamma_{1,2 \to 3}$ can be related to the splitting function $P_{3 \to 1}(\frac{x_1}{x_3})$ which appears in the A-P equations. We have

$$\Gamma_{1,2 \to 3}(x_1, x_2, x_3) = C_{1,2 \to 3} \frac{x_1 x_2}{x_3^2} P_{3 \to 1}(\frac{x_1}{x_3}) \qquad (8.39)$$

where $C_{1,2 \to 3}$ is just $g^2 (= 4\pi\alpha_s) \times$ a colour factor ($\frac{1}{6}$ for $qG \to q$, $\frac{4}{9}$ for $q\bar{q} \to G$, $\frac{3}{8}$ for $GG \to G$).

The term $T^{(2)}$ in (8.36) is the two-parton density which Mueller and Qiu (1986) approximate by

$$T^{(2)}(x_1, x_2, Q^2) = \frac{3}{2} R_A \; \bar{n} \; p_1(x_1, Q^2) \; p_2(x_2, Q^2) \qquad (8.40)$$

where $\bar{n}$ is the nucleon density and $p_i(x, Q^2)$ the single parton densities. If we think of the spatial distribution of partons in the proton where the valence quarks are centrally placed, with a disperse distribution of sea quarks and gluons then the likelihood of a valence quark 'leaking' into a neighbouring nucleon is far smaller than for a gluon. So we can estimate the contribution of recombination to the valence quark distribution as just the

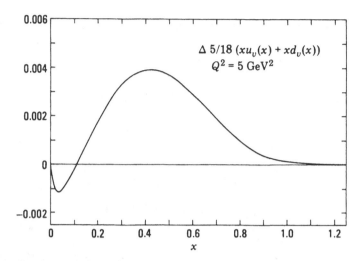

Fig. 8.11   The contribution to the valence quark distribution due to gluons leaking into neighbouring nucleons.

contribution of fig. 8.9(a) which is

$$
x\Delta q_v(x) = \frac{3}{2} R_A \bar{n} \frac{x}{Q^2} \int dx_1 dx_2 q_v(x_1) G(x_2)
$$
$$
\times \Gamma_{qG \to q}(x_1, x_2, x_1 + x_2)[\delta(x - x_1 - x_2) - \delta(x - x_1)]
$$
$$
= \frac{3}{4} \frac{A^{1/3} \alpha_s}{R^2 Q^2} \int dx_1 dx_2 x_1 q_v(x_1) G(x_2) \frac{x}{(x_1 + x_2)}
$$
$$
\times \left[1 + (\frac{x_1}{x_1 + x_2})^2\right][\delta(x - x_1 - x_2) - \delta(x - x_1)] \quad (8.41)
$$

where R is the radius of the proton. The combination of δ-functions plays the same role as in (5.33): it serves to cancel the singularity in the gluon density as $x_2 \to 0$. The δ-functions also guarantee that $\int_0^1 dx \Delta q_v(x) = 0$, i.e. that the *number* of valence quarks does not alter.

Fig. 8.11 shows the modification to the valence quark distribution where one can see explicitly the distribution extending beyond $x = 1$. From the figure it is clear that the valence quarks have *gained* momentum.

It is then straightforward to compute the modification to the structure function $\Delta F_2(x, Q^2)$ by including the contributions from graphs (b), (c) of fig. 8.9. Next, one can calculate the modification to the gluon density from graphs (b), (c) and (d) but the last one almost completely dominates – i.e. the fusion of two gluons to form one gluon, thereby suppressing the gluon density. This contribution to the gluon distribution is given by

$$
\begin{aligned}
x\Delta G(x) &= \frac{3}{2} R_A \bar{n} \frac{x}{Q^2} \int dx_1 dx_2 G(x_1) G(x_2) \\
&\quad \times \Gamma_{GG \to G}(x_1, x_2, x_1 + x_2) \\
&\quad \times [\delta(x - x_1 - x_2) - \delta(x - x_1) - \delta(x - x_2)] \\
&= \frac{27}{16} \frac{A^{1/3} \alpha_s}{R^2 Q^2} \int dx_1 dx_2 G(x_1) G(x_2) \frac{1}{(x_1 + x_2)^4} \\
&\quad \times [(x_1 x_2)^2 (x_1^2 + x_2^2) + x_1^2 x_2^2] \\
&\quad \times [\delta(x - x_1 - x_2) - \delta(x - x_1) - \delta(x - x_2)] \quad (8.42)
\end{aligned}
$$

Again the δ-functions take care of the singularities in the gluon densities. In this case they ensure momentum conservation for this contribution alone. The extra contributions from graphs (b) and (c) guarantee overall momentum conservation between the gluons, quarks and antiquarks. Fig. 8.12 shows the resulting modification of the gluon distribution due to parton fusion. At low x one sees the dramatic shadowing of the gluons due to gluon recombination. Shadowing disappears by the time x reaches 0.08 and for larger x values the gluon distribution itself is already rapidly decreasing.

8.7.2 Q^2-dependence and shadowing of the structure function

Mueller and Qiu (1986) showed that the evolution equations for the modified quark and gluon distributions are themselves modified by parton recombination effects. These correction terms to the $\ln Q^2$ derivative also have a factor $1/Q^2$. Therefore the quark distributions in a nucleus and in a free nucleon will have the same derivative at large Q^2. So if the sea-quarks are suppressed in the nucleus for some relatively low value of Q^2, they will continue to be so at higher Q^2 because the slopes of the nuclear quark distributions are always smaller than the proton quark distributions, i.e. shadowing persists to relatively large Q^2 (Qiu 1987).

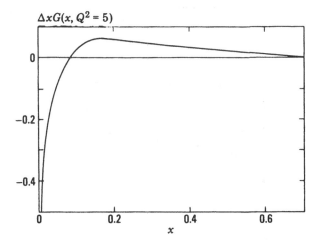

Fig. 8.12   The ratio of the modified gluon distribution, as a result of gluon recombination, to the unmodified distribution for $Q^2 = 5$ GeV^2.

As $x \to 0$, the behaviour of the quark distribution is governed by the behaviour of the gluons in that region. This was discussed in detail in section 6.4 where we saw that the rapid build up of the gluon density at low x was responsible, through the A-P equations, for the steep rise in the sea-quarks at low x. The link between the shapes of each distribution in this region was given by (6.37) and we may expect a similar relation to hold approximately for the shadowing modifications, i.e.

$$\delta x q(x, Q^2) \simeq -\frac{x}{12} \frac{\partial \Delta x G(x, Q^2)}{\partial x} \qquad (8.43)$$

To get a rough estimate of how gluon shadowing translates into shadowing of the sea-quarks we use (8.43) although we cannot expect the relation to be strictly valid for the modified distributions. Fig. 8.13 shows the result for the shadowed quark distributions at low x when the modification from (8.43) is included.

We see that the size of the suppression for the quarks is greater than for the gluons, i.e. shadowing actually gets magnified as it translates from the gluons to the structure function. This fact can be seen in the exact expression at small x for the evolution of the ratio of structure functions. $R_F = F_2^A(x, Q^2)/F_2(x, Q^2)$

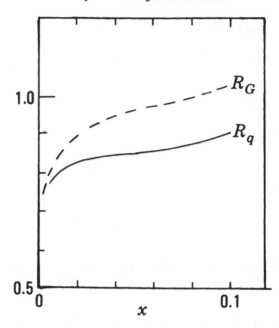

Fig. 8.13   The ratio of the effective nuclear and nucleon parton distributions at $Q^2 = 5$ GeV^2. The dashed line is $R_G = 1 + \Delta xG/xG$ of fig. 8.12 and the solid line is $R_q = 1 + \Delta xq/xq + \delta xq/xq$ computed from the graphs of fig. 8.9 and the correction of (8.43).

(assuming F_2 is dominated by sea-quarks) which involves $R_G = xG^A(x,Q^2)/xG(x,Q^2)$, the ratio shown in fig. 8.12. The evolution of R_F can be expressed (Qiu 1987) as

$$\frac{\partial R_F}{\partial \ln Q^2} \sim \frac{1}{xq_s} \left[\int_x^1 \frac{dy}{y} \frac{x}{y} P_{qG}(\frac{x}{y}) yG(y,Q^2) \right.$$

$$\times \{ R_G(x,Q^2) - R_F(x,Q^2) \} - \left(\frac{3}{2} R_A \bar{n} - \frac{9}{8\pi R^2} \right) \frac{1}{Q^2}$$

$$\left. \times \text{term} \propto \alpha_s (xG(x,Q^2))^2 \right] \qquad (8.44)$$

Lowering the value of R_G makes the derivative on the l.h.s. smaller and so allows the shadowing of the structure function to persist to moderately large Q^2. Furthermore, the eventual fading away of the shadowing requires the derivative to be positive which, from (8.44), then requires $R_G > R_F$, i.e. shadowing for the structure function at low x is bigger than the gluon shadowing.

To summarise, gluon recombination is a mechanism within the framework of perturbative QCD which can qualitatively account for the observed x and Q^2 dependences of the shadowing of the structure function. The data at very low x is still measured only at low Q^2, where VDM models are quantitatively successful (Badelek and Kwiecinski 1988, 1989). It will therefore be interesting to see whether shadowing (at low x) persists to higher Q^2. Recombination of partons is therefore a natural mechanism to add to the existing perturbative QCD modification of parton distributions given by Q^2 rescaling. Phenomenologically, parton recombination improves the the description of the nuclear structure functions at small and large x, so a complete description of nuclear structure functions at large Q^2 entirely within the context of perturbative QCD is possible. Quantitative calculations of the combined mechanisms have yet to be carried out.

Appendix

Radiative corrections

One of the headaches encountered in extracting proton structure functions from DIS arises from the fact that, in addition to the lowest order graph of fig. 2.1, there are $O(\alpha^3)$ processes which contribute to the cross-section. In fact there are a whole set of electroweak radiative processes which must be computed in order to extract the Born contribution of fig. 2.1. At HERA such radiative corrections can be very large and it is essential to have some estimate of *where* these corrections are important and how reliably they can be computed.

In this appendix we simply sketch out the problems involved and present the results of published computations. A general overview is given by Hollik (1987) where the whole procedure is elegantly summarised. We consider just the QED corrections to begin with.

A.1 Radiation from the electron

Consider the emission of a real photon from the incoming or final electron, as shown in fig. A.1.

These contributions were first discussed by Mo and Tsai (1969) who recognised these as the source of the largest corrections. In addition to the usual variables s, Q^2 and W^2 given by

$$s = (k+p)^2 \qquad (k-k')^2 = -Q^2 \qquad W^2 = (p_n + l)^2 \qquad (A.1)$$

we introduce variables s', Q'^2, W'^2 given by

$$s' = (k'+l)^2 \qquad (p-p_n)^2 = -Q'^2 \qquad W'^2 = p_n^2 \qquad (A.2)$$

The cross-section for this real photon emission is then

$$d\sigma_R = \frac{4M\alpha^3}{\pi^2(s-M^2)Q'^4} \, \ell^{\mu\nu} W_{\mu\nu} dR_{n+2} \qquad (A.3)$$

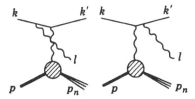

Fig. A.1   Simplest $O(\alpha^3)$ graphs for real photon emission at the leptonic vertex.

where $\ell^{\mu\nu}, W_{\mu\nu}$ are the familiar leptonic and hadronic tensors (see chapter 2) and R_{n+2} is the $(n+2)$ particle phase space given by

$$dR_{n+2} = \frac{\pi}{16\sqrt{\lambda(s, m^2, M^2)(-\Delta_4)}}\, dQ^2 dQ'^2 ds'\, dW^2 dW'^2\, dR_n$$

$$(A.4)$$

This means we can express σ_R in terms of the familiar structure functions W_1 and W_2,

$$\frac{d\sigma_R}{dW^2 dQ^2} = \frac{2\alpha^3}{(s-M^2)^2}\int \frac{dW'^2 dQ'^2}{Q'^4}\, [2MW_1(W'^2, Q'^2)\, S_1$$
$$+ \frac{1}{2M} W_2(W'^2, Q'^2)\, S_2]$$

$$(A.5)$$

Here S_1 and S_2 are complicated functions of W'^2, W^2, Q'^2, Q^2 (see Akhundov, Bardin and Shumeiko 1977) and contain factors $(Q^2 - Q'^2)^{-1}$, so as $l \to 0$ the cross-section diverges.

Equation (A.5) illustrates the nature of the radiative correction problem, namely the intimacy between the structure functions of the proton and the electromagnetic interaction between quarks and leptons. The *hadronic* part cannot be computed from first principles and has to be treated as experimental *input* to the radiative correction. On the other hand this hadronic input cannot be extracted from experiment without knowledge of the radiative correction. In practice the analysis of the measured cross-section involves an iteration procedure starting from an educated guess for the hadronic input which then converges to a refined output after several steps.

Of course the divergence from the real photon emission of fig. A.1 is spurious; it is cancelled by corresponding divergence associated with virtual photons in fig. A.2. Nevertheless, the contributions arising from real photons radiated from the electron

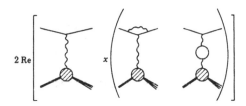

Fig. A.2 $O(\alpha^3)$ contributions from virtual photons associated with the leptonic vertex.

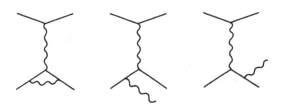

Fig. A.3   QED corrections to a quark in the proton.

and the leptonic vertex correction account for the bulk of all radiative corrections.   Large logarithmic terms of the form $(\alpha/\pi)\ln(Q^2/m_e^2)$ are typically associated with the radiation of collinear photons.   Since there is a $1/Q^4$ factor in the cross-section , the emission of a very energetic photon which shifts the value to $1/Q\prime^2$ (smaller than that determined from the energy and momentum of the final electron) produces a large change in the cross section. From the expression for y in terms of the energies E, E' of the incoming and outgoing electrons we conclude that a hard collinear photon prefers large y (and small x for fixed Q^2) while soft photon radiation prefers small y (large x).

A.2   Radiation from the hadron

Next we must include corrections arising from photons radiated from hadronic vertex and here we simply consider the QED corrections associated with a quark, fig. A.3.

These corrections are far less important and clearly do not lead to a 'misidentification' of Q^2 which occurs when photons radiate off the electrons. The problem now is that we get terms of the form $(\alpha/\pi)\ln(Q^2/m_q^2)$, where m_q is the quark mass, arising from photons collinear with the quark. Such mass singularities should

Fig. A.4   Interference between radiation of photons from the lepton and quark lines.

be regarded as an integral part of the quark distributions $xq(x, Q^2)$ however. Correctly, one should define the quark density $q(x, Q^2)$ in terms of a QED corrected 'bare' density q^0,

$$q(x, Q^2) = q^0(x)(1 + \delta_h(Q^2)) \qquad (A.6)$$

where $\delta_h(Q^2)$ is the *hadronic* part of the QED corrections. By absorbing the singular leading logarithmic contribution in this way, the density at one value of Q^2 is related to that at Q_0^2 by

$$q(x, Q^2) = q(x, Q_0^2)[1 + \delta_h(Q^2) - \delta_h(Q_0^2)] \qquad (A.7)$$

and now the leading log corrections are of the form $(\alpha/\pi)\ln(Q^2/Q_0^2)$ and quite harmless. In fact this QED correction should explicitly appear in the AP equations (Kripfganz and Perlt 1987) for the evolution of $q(x, Q^2)$.

Finally the interference between the photons associated with leptonic hadronic vertices should be included, fig. A.4.

These corrections cannot be absorbed into the quark densities; the terms which appear are typically logs or dilogs with argument (Q^2/xs).

A.3   Results

To display the results of calculating the QED corrections one computes the relative correction to the Born term of fig. 2.1,

$$\delta_a = \frac{\mathrm{d}^2\sigma^a}{\mathrm{d}x\,\mathrm{d}y} \Big/ \frac{\mathrm{d}^2\sigma^{Born}}{\mathrm{d}x\,\mathrm{d}y} \qquad (A.8)$$

where a labels the individual contributions from photons radiated from the lepton lines, quark lines and interference. Fig. A.5 shows the results for $a = e, q, i$ calculated by Bardin, Burdik, Christova and Riemann (1987).

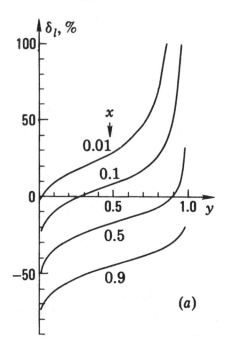

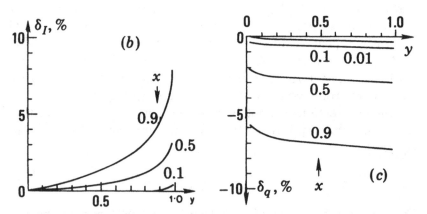

Fig. A.5   QED radiative corrections from photon exchange due to radiation
from (a) lepton lines, (c) quark lines and (b) their interference for the neutral
current interaction at HERA for $s = 10^5$ GeV^2 (Bardin *et al.* 1987.

Finally one should include the corrections from weak interactions. This is essentially a straightforward procedure with no inherent problems. The full electroweak corrections have been computed by Bohm and Spiesberger (1987) and Bardin, Burdik, Christova and (1987). The charged current corrections have been evaluated by these two groups. In this case, a separation into the QED and weak corrections is not sensible since it is gauge dependent procedure. Here the total corrections to the cross-section are below 10% for $Q^2 < 10^5$ GeV^2 and even below 5% for $Q^2 < 10^4$ GeV^2.

References

Abramowicz, H. *et al.* (1983). , *Zeit. Phys.* **C17**, 283.
Adler, S.L. (1966). , *Phys. Rev.* **143**, 1144.
Adler, S.L. (1969). , *Phys. Rev.* **177**, 2426.
Akesson, T. *et al.* (1987). , *Zeit. Phys.* **C34**, 293.
Akulinichev, S.V., Kulagin, S.A. and Vagradov, G.M. (1985). , *Phys. Lett.* **158B**, 485.
Akulinichev, Shlomo, S., S.V., Kulagin, S.A. and Vagradov, G.M. (1985). , *Phys. Rev. Lett.* **55**, 2239.
Akhundov, A.A., Bardin, D.Yu. and Shumeiko, N.M. (1977). , *Sov. J. Nucl. Phys.* **26**, 660.
Altarelli, G. (1982). , *Phys. Rep.* **81**, 1.
Altarelli, G. and Parisi, G. (1977). , *Nucl. Phys.* **B126**, 298.
Altarelli, G. and Ross, G.G. (1988). , *Phys. Lett.* **212B**, 391.
Altarelli, G. and Stirling, W.J. (1989). , *Part. World* **1**, 40.
Altarelli, G., Ellis, R.K. and Martinelli, G. (1978). , *Nucl. Phys.* **B143**, 521; **B146**, 544(E).
Altarelli, G., Ellis, R.K. and Martinelli, G. (1979). , *Nucl. Phys.* **B147**, 461.
Amati, D., Bassetto, A., Ciafaloni, M., Marchesini, G. and Veneziano, G. (1980). , *Nucl. Phys.* **B173**, 429.
Appel, J.A. *et al.* (1985). , *Phys. Lett.* **160B**, 349.
Arneodo, M. *et al.* (1988). , *Phys. Lett.* **211B**, 493.
Arnold, R.G. *et al.* (1984). , *Phys. Rev. Lett.* **52**, 727.
Ashman, J. *et al.* (1988a). , *Phys. Lett.* **206B**, 364.
Ashman, J. *et al.* (1988b). , *Phys. Lett.* **202B**, 603.
Ashman, J. *et al.* (1989). , *Nucl. Phys.* **B328**, 1.
Aubert, J.J. *et al.* (1982). , *Phys. Lett.* **123B**, 275.
Aubert, J.J. *et al.* (1983). , *Nucl. Phys.* **B213**, 31.
Aubert, J.J. *et al.* (1985). , *Nucl. Phys.* **B259**, 189.
Aubert, J.J. *et al.* (1986). , *Nucl. Phys.* **B272**, 158.
Aubert, J.J. *et al.* (1987). , *Nucl. Phys.* **B293**, 740.
Aurenche, P., Baier, R., Fontannaz, M. and Schiff, D. (1986). , *Phys. Lett.* **169B**, 441.
Aurenche, P., Baier, R., Fontannaz, M. and Schiff, D. (1988). , *Nucl. Phys.* **B297**, 661.
Aurenche, P., Baier, R., Douri, A., Fontannaz, M. and Schiff, D. (1984). , *Phys. Lett.* **140B**, 87.
Badelek, B. and Kwiecinski, J. (1988). , *Phys. Lett.* **208B**, 508.
Badelek, B. and Kwiecinski, J. (1989). , *Zeit. Phys.* **43**, 251.
Balitzkii, Y.Y. and Lipatov, L.N. (1978). , *Yad. Fiz.* **28**, 585.

Bardeen, W.A., Buras, A.J., Duke, D.W. and Muta, T. (1978). , *Phys. Rev.* **D18**, 3998.

Bardin, D.Yu., Burdik, C., Christova, P.Ch. and Riemann, T. (1987). Dubna preprint E2-87-595.

Bardin, D.Yu., Burdik, C., Christova, P.Ch. and Riemann, T. (1988). , *Zeit. Phys.* **C42**, 679.

Bari, G., *et al.* (1985). , *Phys. Lett.* **163B**, 282.

Bell, J.S. and Jackiw, R. (1969). , *Nuo. Cim.* **A51**, 47.

Benvenuti, A.C. *et al.* (1987). , *Phys. Lett.* **195B**, 91.

Benvenuti, A.C. *et al.* (1989). , *Phys. Lett.* **223B**, 485.

Berge, J.P. *et al.* (1989). CERN preprint EP/89-103.

Berger, E.L., Coester, F. and Wiringa, R.B. (1984). , *Phys. Rev.* **D29**, 398.

Bickerstaff, R.P. and Thomas, A.W. (1987). , *Phys. Rev.* **D35**, 108.

Bickerstaff, R.P. and Thomas, A.W. (1989). , *J. Phys.* **G15**, 1523.

Bitar, K., Johnson, P.W. and Tung, W.K. (1979). , *Phys. Lett.* **33**, 244.

Bjorken, J.D. (1966). , *Phys. Rev.* **148**, 1476.

Bjorken, J.D. (1967). , *Phys. Rev.* **163**, 1767.

Bjorken, J.D. (1969). , *Phys. Rev.* **179**, 1547.

Bjorken, J.D. (1970). , *Phys. Rev.* **D1**, 1376.

Bloom, E.D. *et al.* (1969). , *Phys. Rev. Lett.* **23**, 930.

Bloom, E.D. and Gilman, F. (1971). , *Phys. Rev.* **D4**, 2901.

Bodek, A. *et al.* (1979). , *Phys. Rev.* **D20**, 1471.

Böhm, M. and Spiesberger, H. (1987). , *Nucl. Phys.* **B294**, 108.

Bonesini, M. *et al.* (1987). , *Zeit. Phys.* **C37**, 39,535.

Brandt, R.A. and Preparata, G. (1971). , *Nucl. Phys.* **B27**, 541.

Braun,V.M. and Kolesnichenko, A.V. (1987). , *Nucl. Phys.* **B283**, 723.

Breidenbach, M. *et al.* (1969). , *Phys. Rev. Lett.* **23**, 935.

Brock, R. (1980). , *Phys. Rev. Lett.* **44**, 1027.

Brodsky, S.J. and Farrar, G. (1973). , *Phys. Rev. Lett.* **31**, 1153.

Brodsky, S.J. and Farrar, G. (1975). , *Phys. Rev.* **D11**, 1309.

Brodsky, S.J. and Lepage, G.P. (1979). SLAC Summer Institute SLAC-REP.224.

Brodsky, S.J. and Lepage, G.P. (1980). , *Phys. Rev.* **D22**, 2157.

Brodsky, S.J. and Lepage, G.P. (1981). , *Phys. Rev.* **D24**, 2848.

Brown C.N. *et al.* (1989). See review talk by J. Huston, XIVth Symposium on Lepton-Photon Interactions, Stanford.

Brown C.N. *et al.* (1990). , *Phys. Rev. Lett.* **63**, 2637.

Buck, B. and Perez, S.M. (1983). , *Phys. Rev. Lett.* **50**, 1975.

Burkhadt, H. and Cottingham, W.N. (1970). , *Ann. Phys.* **56**, 453.

Cahn, R.N. (1977). , *Phys. Lett.* **78B**, 269.

Callan, C.G. (1970). , *Phys. Rev.* **D2**, 1541.

Callan, C.G. and Gross D.J. (1969). , *Phys. Rev. Lett.* **22**, 156.

Caswell, W. (1974). , *Phys. Rev. Lett.* **33**, 244.

Chernyak, V.L. and Zhitnitsky, A.R. (1984). , *Phys. Rep.* **112**, 173.

Clark, A.R. *et al.* (1980). , *Phys. Rev.* **45**, 1465.

Close, F.E. (1978). Structure Functions and Counting Rules, in *19th International Conference on High Energy Physics*, ed. S. Homma, M. Kawaguchi, H. Miyazawa (Phys. Soc. of Japan, Tokyo).

Close, F.E. (1979). *An Introduction to Quarks and Partons* (Academic Press, New York).

Close, F.E. and Sivers, D. (1979). , *Phys. Rev. Lett.* **39**, 1116.

Close, F.E., Halzen, F. and Scott, D.M. (1977). , *Phys. Lett.* **68B,** 447.

Close, F.E. Roberts, R.G. and Qiu, J. (1989). , *Phys. Rev.* **40,** 2820.

Close, F.E., Roberts, R.G. and Ross, G.G. (1983). , *Phys. Lett.* **129B,** 346.

Close, F.E., Roberts, R.G. and Ross, G.G. (1988). , *Nucl. Phys.* **B296,** 582.

Close, F.E., Jaffe, R.L., Roberts, R.G. and Ross, G.G. (1985). , *Phys. Rev.* **D13,** 1004.

Collins ,J.C. (1985). QCD Results for small x, in *Supercollider Physics, Oregon Workshop on Super High Energy Physics, Eugene, Oregon,* ed. D. Soper (World Scientific, Singapore).

Collins, J.C. and Kwiecinski, J. (1988). , *Nucl. Phys.* **B316,** 307.

Collins, J.C. and Qiu, J.W. (1988). Argonne preprint.

Collins, J.C. and Tung, W.K. (1986). , *Nucl. Phys.* **B278,** 934.

Cooper-Sarkar, A.M., Ingelman, G., Long, K.R., Roberts, R.G. and Saxon, D.H. (1988). , *Zeit. Phys.* **C39,** 281.

Curci ,G., Furmanski, W. and Petronzio, R. (1980). , *Nucl. Phys.* **B175,** 27.

Dasu, S. *et al.* (1988). , *Phys. Rev. Lett.* **60,** 2591.

De Rujula, A., Georgi, H. and Politzer, H.D. (1977). , *Ann. Phys.* **1103,** 315.

Dokshitzer, Yu.L., Dyakonov, D.I. and Troyan, S.I. (1978). , *Phys. Lett.* **78B,** 290.

Dokshitzer, Yu.L., Dyakonov, D.I. and Troyan, S.I. (1980). , *Phys. Rep.* **58,** 269.

Drell, S.D. and Yan, T.M. (1970). , *Phys. Rev. Lett.* **24,** 181.

Drell, S.D. and Yan, T.M. (1971). , *Ann. Phys.* **66,** 595.

Duke, D.W. and Owens, J.F. (1984). , *Phys. Rev.* **D30,** 49.

Duke, D.W. and Roberts, R.G. (1985). , *Phys. Rep.* **120,** 275.

Duke, D.W., Kimel, J.D. and Sowell, G.A. (1982). , *Phys. Rev.* **D25,** 71.

Duke, D.W., Owens, J.F. and Roberts, R.G. (1982). , *Nucl. Phys.* **B195,** 285.

Ellis, J. and Jaffe, R.L. (1974). , *Phys. Rev.* **D9,** 1444; addendum **D10,**1669.

Ellis, R.K., Furmanski, W. and Petronzio, R. (1980). , *Nucl. Phys.* **B212,** 29.

Ericson, M. and Thomas, A.W. (1983). , *Phys. Lett.* **B128,** 112.

Fajfer, S. and Oakes, R.J. (1986). , *Phys. Lett.* **176B,** 473.

Farrar, G.R. and Jackson, D.R. (1975). , *Phys. Rev. Lett.* **35,** 1416.

Feynman, R.P. (1969). , *Phys. Rev. Lett.* **23,** 1415.

Feynman, R.P. (1972). *Photon-Hadron Interactions* (W.A. Benjamin, NewYork).

Floratos, E.G., Kounnas, C. and Lacaze, R. (1981). , *Nucl. Phys.* **B192,** 417.

Floratos, E.G., Ross, D.A. and Sachrajda, C.T. (1977). , *Nucl. Phys.* **B129,** 66; **B139,**545(E).

Floratos, E.G., Ross, D.A. and Sachrajda, C.T. (1979). , *Nucl. Phys.* **B152,** 493.

Foudas, C. *et al.* (1988). Rochester preprint UR-1057, Fermilab-Conf-88/160-E.

Frankfurt, L.L. and Strikman M.I. (1987). , *Phys. Lett.* **183B,** 254.

Frankfurt, L.L. and Strikman M.I. (1988). , *Phys. Rep.* **160,** 235.

Gell-Mann, M. and Low, F.E. (1954). , *Phys. Rev.* **95,** 1300.

Georgi, H. and Politzer, H.D. (1974). , *Phys. Rev.* **D9,** 416.

Georgi, H. and Politzer, H.D. (1976). , *Phys. Rev.* **D14,** 1829.

Glück, M., Hoffman E. and Reya, E. (1982). , *Zeit. Phys.* **C13,** 119.

Gonzalez-Arroyo, A., Lopez C. and Yndurain, F.J. (1979). , *Nucl. Phys.* **B153,** 161.

Gonzalez-Arroyo, A., Lopez C. and Yndurain, F.J. (1980). , *Nucl. Phys.* **B166**, 429.

Gordon, B.A. *et al.* (1979). , *Phys. Rev.* **D20**, 2645.

Gottfried, K. (1967). , *Phys. Rev. Lett.* **18**, 1154.

Gribov, L.V., Levin, E.M. and Ryskin, M.G. (1983). , *Phys. Rep.* **100**, 1.

Gross, D.J. and Llewellyn Smith, C.H. (1969). , *Nucl. Phys.* **B14**, 337.

Gross, D.J. and Treiman, S.B. (1971). , *Phys. Rev.* **D4**, 1059.

Gross, D.J. and Wilczek, F.A. (1973). , *Phys. Rev. Lett.* **30**, 1343.

Gross, D.J. and Wilczek, F.A. (1974). , *Phys. Rev.* **D9**, 980.

Gross, D.J., Treiman, S.B. and Wilczek, F.A. (1977). , *Phys. Rev.* **D15**, 2486.

Harriman, P.N., Martin, A.D., Stirling, W.J. and Roberts, R.G. (1990). RAL report 90-007.

Herrod, R.T. and Wada, S. (1980). , *Phys. Lett.* **96B**, 195.

Herrod, R.T., Wada, S. and Webber, B.R. (1981). , *Zeit. Phys.* **C9**, 351.

Hollik, W. (1987). Radiative corrections in deep inelastic scattering, in *Proceedings of the HERA Workshop, p 579*, ed. R. Peccei (DESY, Hamburg).

Iijima, B. (1982). MIT preprint CTP#993.

Jackson, J.D., Ross, G.G. and Roberts, R.G. (1989). , *Phys. Lett.* **226B**, 159.

Jaffe, R.L. (1985). Deep inelastic scattering with application to nuclear targets , in *Los Alamos School on Relativistic Dynamics and Quark-Nuclear Physics*, ed. M.B. Jackson and A. Picklesimer. (John Wiley and Sons, New York).

Jaffe, R.L., Close, F.E., Roberts, R.G. and Ross, G.G. (1984). , *Phys. Lett.* **134B**, 449.

Jaroszewicz, T. (1980). , *Act. Phys. Polon.* **B11**, 965.

Jones, D.R.T. (1974). , *Nucl. Phys.* **B75**, 531.

Kodaira, J., Matsuda, S., Sasaki, K. and Uematsu, T. (1979*b*). , *Nucl. Phys.* **B159**, 99.

Kodaira, J., Matsuda, S., Muta, T., Sasaki, K. and Uematsu, T. (1979*a*). , *Phys. Rev.* **D20**, 627.

Kogut, J. and Susskind, L. (1974). , *Phys. Rev.* **D9**, 697,706,3391.

Koltun, D.S. (1972). , *Phys. Rev. Lett.* **28**, 182.

Kripfganz, J. and Perlt, H. (1987). Electroweak radiative corrections and quark mass singularities, in *Proceedings of the HERA Workshop, p645* , ed. R. Peccei (DESY, Hamburg).

Kuraev, E.A., Lipatov, L.N. and Fadin, V.S. (1977). , *Z.E.T.P.* **72**, 377.

Kwiecinski, J. (1985). , *Zeit. Phys.* **C29**, 147,651.

Landshoff. P.V. (1977). , *Phys. Lett.* **66B**, 452.

Landshoff, P.V. and Polkinghorne, J.C. (1972). , *Phys. Rep.* **5c**, 1.

Li, G.L., Liu, K.F. and Brown, G.E. (1988). , *Phys. Lett.* **213B**, 531.

Lipatov, L.N. (1985). Leningrad Nuclear Physics Inst. preprint 1137.

Llewellyn Smith, C.H. (1983). , *Phys. Lett.* **B128**, 107.

Marciano, W.J. (1984). , *Phys. Rev.* **D29**, 5801.

Martin, A.D., Ng, C.K. and Stirling, W.J. (1987). , *Phys. Lett.* **191B**, 200.

Martin, A.D., Roberts, R.G. and Stirling, W.J. (1988*a*). , *Phys. Rev.* **D37**, 1161.

Martin, A.D., Roberts, R.G. and Stirling, W.J. (1988*b*). , *Phys. Lett.* **B206**, 327.

Martin, A.D., Roberts, R.G. and Stirling, W.J. (1989*a*). , *Mod. Phys. Lett.* **A4**, 1135.

Martin, A.D. Roberts, R.G. and Stirling, W.J. (1989*b*). , *Phys. Lett.* **228B**, 149.

Martinelli, G. and Sachrajda, C. (1989). , *Nucl. Phys.* **B316**, 355.

Matveev, V.A., Murddyan, R.M. and Tavkheldize, A.N. (1973). , *Lett. Nuo. Cim.* **7**, 719.

Mestayer, M.D. *et al.* (1983). , *Phys. Rev.* **D27**, 285.

Miramontes, J.L. (1985). Ph.D. thesis, University of Santiago.

Miramontes, J.L. and Sanchez Guillen, J. (1988). , *Zeit. Phys.* **C41**, 247.

Miramontes, J.L., Miramontes, M.A. and Sanchez Guillen, J. (1989). , *Phys. Rev.* **D40**, 2184.

Mo, L.W. and Tsai, Y.S. (1969). , *Rev. Mod. Phys.* **41**, 205.

Moniz, E.J. *et al.* (1971). , *Phys. Rev. Lett.* **25**, 445.

Mueller, A. and Qiu, J. (1986). , *Nucl. Phys.* **B268**, 427.

Nachtmann, O. (1973). , *Nucl. Phys.* **B63**, 237.

Panofsky, W.K.H. (1968). Rapporteurs talk , in *International Conference on High Energy Physics, Vienna,* ed. J. Prentki and J. Steinberger (CERN, Geneva).

Paschos, E. and Wolfenstein, L. (1973). , *Phys. Rev.* **D7**, 91.

Pennington, M.R. (1982). , *Phys. Rev.* **D26**, 2048.

Pennington, M.R. (1983). , *Rep. Prog. Phys.* **46**, 393.

Pennington, M.R. and Ross, G.G. (1979). , *Phys. Lett.* **186B**, 371.

Pennington, M.R. and Ross, G.G. (1981). , *Phys. Lett.* **102B**, 167.

Politzer, H.D. (1973). , *Phys. Rev. Lett.* **30**, 1346.

Politzer, H.D. (1980). , *Nucl. Phys.* **B172**, 349.

Poucher, J.S. *et al.* (1974). , *Phys. Rev. Lett.* **32**, 118.

Qiu, J. (1987). , *Nucl. Phys.* **B291**, 746.

Ratcliffe, P. (1987). , *Phys. Lett.* **192B**, 180.

Renton, P.B. (1986). IX Warsaw Symposium on Elementary Physics, Oxford preprint, 63/86.

Ross, G.G. (1989). review talk at the XIVth Symposium on Lepton-Photon Interactions, Stanford..

Sivers, D. (1982). , *Ann. Rev. Part. Sci.* **32**, 149.

Sloan, T., Smadja, G. and Voss, R. (1988). , *Phys. Rep.* **162**, 45.

Stevenson, P.M. (1981*a*). , *Phys. Lett.* **100B**, 161.

Stevenson, P.M. (1981*b*). , *Phys. Rev.* **D23**, 2916.

Stirling,W.J.S. (1987). QCD at short distances, in *Int. Symposium on Lepton and Photon Interactions, Hamburg,* ed. W. Bartel and R. Rückl (North-Holland, Amsterdam).

Stückelberg, E.C.G. and Peterman, A. (1953). , *Helv. Phys. Acta* **26**, 499.

Symanzik, K. (1970). , *Comm. Math. Phys.* **18**, 227.

't Hooft, G. (1972). unpublished.

't Hooft, G. and Veltman,M. (1972). , *Nucl. Phys.* **B44**, 189.

Tkaczyk, S.M., Stirling, W.J.S. and Saxon, D.M. (1988). Rutherford preprint RAL-88041.

Vainshtein, A.I. and Zakharov, V.I. (1978). , *Phys. Lett.* **72B**, 368.

Varvell, K. *et al.* (1987). , *Zeit. Phys.* **C36**, 1.

Voss, R. (1987). Charged Lepton Interactions, in *Int. Symposium on Lepton and Photon Interactions, p 581,* ed. W. Bartel and R. Rückl (North-Holland, Amsterdam).

Wandzura, S. and Wilczek, F. (1977). , *Phys. Lett.* **72B**, 195.

West, G.B. (1970). , *Phys. Rev. Lett.* **24**, 1206.

Whitlow, L.W. *et al.* (1989). SLAC preprint, SLAC-PUB-5100.
Wilson, K.G. (1969). , *Phys. Rev.* **179**, 1499.
Yoshino, T. and Hagiwara, K. (1984). , *Zeit. Phys.* **C24**, 185.
Zee, A., Wilczek, F. and Treiman, S.B. (1974). , *Phys. Rev.* **D10**, 2881.

Index

Printed in the United States
By Bookmasters